International Assessment of Research and Development in Simulation-Based Engineering and Science

International Assessment of Research and Development in Simulation-Based Engineering and Science

Editor

Sharon C Glotzer

University of Michigan, Ann Arbor, USA

ICP

Imperial College Press

Published by

Imperial College Press
57 Shelton Street
Covent Garden
London WC2H 9HE

Distributed by

World Scientific Publishing Co. Pte. Ltd.
5 Toh Tuck Link, Singapore 596224
USA office: 27 Warren Street, Suite 401-402, Hackensack, NJ 07601
UK office: 57 Shelton Street, Covent Garden, London WC2H 9HE

British Library Cataloguing-in-Publication Data
A catalogue record for this book is available from the British Library.

This document was sponsored by the National Science Foundation under a cooperative agreement from NSF (ENG-0844639) to the World Technology Evaluation Center, Inc. The U.S. Government retains a nonexclusive and nontransferable license to exercise all exclusive rights provided by copyright. Any writings, opinions, findings, and conclusions expressed in this material are those of the authors and do not necessarily reflect the views of the U. S. Government, the authors' parent institutions, or WTEC.

INTERNATIONAL ASSESSMENT OF RESEARCH AND DEVELOPMENT IN SIMULATION-BASED ENGINEERING AND SCIENCE

ISBN-13 978-1-84816-697-4
ISBN-10 1-84816-697-4

Typeset by Stallion Press
Email: enquiries@stallionpress.com

Printed in Singapore by B & Jo Enterprise Pte Ltd

DEDICATION

We at WTEC wish to extend our gratitude and appreciation to the panelists for their valuable insights and their dedicated work in conducting this international benchmarking study of R&D in simulation-based engineering and science. We wish also to extend our sincere appreciation to the advisory board, the presenters at the U.S. baseline workshop, and to the panel's site visit hosts for so generously and graciously sharing their time, expertise, and facilities with us. For their sponsorship of this important study, our thanks go to the National Science Foundation (NSF), the Department of Energy (DOE), the National Aeronautics and Space Administration (NASA), the National Institutes of Health (NIH), the National Institute of Standards and Technology (NIST), and the Department of Defense (DOD). We believe this report provides a valuable overview of ongoing R&D efforts in simulation-based engineering and science that can help scientists and policymakers effectively plan and coordinate future efforts in this important field.

R. D. Shelton

CONTENTS

FOREWORD

> We have come to know that our ability to survive and grow as a nation to a very large degree depends upon our scientific progress. Moreover, it is not enough simply to keep abreast of the rest of the world in scientific matters. We must maintain our leadership.[1]

President Harry Truman spoke those words in 1950, in the aftermath of World War II and in the midst of the Cold War. Indeed, the scientific and engineering leadership of the United States and its allies in the twentieth century played key roles in the successful outcomes of both World War II and the Cold War, sparing the world the twin horrors of fascism and totalitarian communism, and fueling the economic prosperity that followed. Today, as the United States and its allies once again find themselves at war, President Truman's words ring as true as they did more than a half-century ago. The goal set out in the Truman Administration of maintaining leadership in science has remained the policy of the U.S. Government to this day. For example, the top goal of the NSF[2] is to, "foster research that will advance the frontiers of knowledge, emphasizing areas of greatest opportunity and potential benefit and establishing the nation as a global leader in fundamental and transformational science and engineering."

The United States needs metrics for measuring its success in meeting this goal. That is one of the reasons that the National Science Foundation (NSF) and many other agencies of the U.S. Government have supported the World Technology Evaluation Center (WTEC) for the past 20 years. While other programs have attempted to measure the international competitiveness of U.S. research by comparing funding amounts, publication statistics, or patent activity, WTEC has been the most significant public domain effort in the U.S. Government to use peer review to evaluate the status of U.S. efforts in comparison to those abroad. Since 1989, WTEC has conducted over 60 such assessments in a wide variety

[1]Remarks by the President on May 10, 1950, on the occasion of the signing of the law that created the National Science Foundation. *Public Papers of the Presidents* 120, 338.
[2]Investing in America's Future: Strategic Plan FY2006-2011, Arlington: NSF 06-48.

of fields including advanced computing, nanoscience and nanotechnology, biotechnology, and advanced manufacturing.

The results have been extremely useful to NSF and other agencies in evaluating ongoing research programs and in setting objectives for the future. WTEC studies also have been important in establishing new lines of communication and identifying opportunities for cooperation between U.S. researchers and their colleagues abroad, thus helping to accelerate the progress of science and technology within the international community. WTEC is an excellent example of cooperation and coordination among the many agencies of the U.S. Government that are involved in funding research and development: almost every WTEC study has been supported by a coalition of agencies with interests related to the particular subject.

As President Truman said over 50 years ago, our very survival depends upon continued leadership in science and technology. WTEC plays a key role in determining whether the United States is meeting that challenge, and in promoting that leadership.

Michael Reischman
Deputy Assistant Director for Engineering
National Science Foundation

PREFACE

In 1946 in Philadelphia, ENIAC, the first large-scale electronic computer in the United States, was constructed and set to work toward calculating the path of artillery shells — a simulation using numerical calculations and based on the equations of physics. It was needed and it was promising, but it was also perceived as a tool too expensive and complicated to reach wide use. Thomas J. Watson, the chairman of IBM, has been widely quoted for remarking in 1943: "I think there is a world market for maybe five computers."

More than 60 years later, we find applications of predictive computer simulation all around us. When storms arise, two- and three-dimensional views are generated, their most likely paths are predicted, and even the uncertainty of the prediction is shown. Uncannily realistic images are conjured for movies, television, and video games. Consumer goods from cars to potato chips are designed, produced, and tested. Guiding wise decision-making, computer simulation is now used to assess stability of our energy networks, safety of industrial processes, collective behavior of people and economies, and global changes in climate.

What changed to bring us to this point? Almost everything:

- Computers were auditorium-sized, multiton facilities based on vacuum tubes, then transistors. They have become diverse in size and use: large supercomputers, multipurpose personal tools, and dedicated, embedded processing units, all based on thumbnail-sized microelectronic chips.
- Computer programs were hard-wired with no means of storage. They have become intricately crafted constructions of logic, created line-by-line or with drag-and-drop logic symbols, stored in memory chips and small rotating disks.
- Because reprogramming required rewiring, input was physical and output was exclusively by printed text. Input now comes from keyboards, mice, sensors, and voice recognition, and results appear on screens, from audio speakers and headsets, and even as three-dimensional objects that are sculpted or deposited by computer control.

- Computers initially were dedicated to a small group of users and were unconnected to each other. The invention of computer networks and then Berners-Lee's invention of the World Wide Web transformed the connectedness of computers and the people using them, as well as commerce, medicine, information access, technology, and intellectual disciplines.

Computer-based simulation has become a crucial part of the present infrastructure, central to applying these advances to the conduct of scientific research and engineering practice. To examine the breadth, depth, and implications of this change, a blue-ribbon panel on "Simulation-Based Engineering Science" was commissioned in 2005, led by Prof. J. Tinsley Oden of the University of Texas at Austin and sponsored by the Engineering Directorate of the National Science Foundation. In their May 2006 report,[1] they proposed SBES as being a new discipline in engineering science, in which modern computational methods, computational thinking, and devices and collateral technologies are combined to address problems far outside the scope of traditional numerical methods. They proposed that advances in SBES offer hope of resolving a variety of fundamental and complex problems in engineering science that affect discovery, health, energy, environment, security, and quality of life of all peoples. At the same time, their view of SBES was of being an interdisciplinary field, lying at the intersection of several disciplines and positioned where many of the important developments of the next century will be made. The basic building blocks of SBES are computational and applied mathematics, engineering science, and computer science, but the fruition of SBES requires the symbiotic development and the enrichment of these fields by a host of developing technologies in data acquisition and management, imaging, sensors, and visualization. Advances in SBES will require advances in all of these component disciplines.

The present, broader study on "Simulation-Based Engineering and Science" (SBE&S) was commissioned to investigate these points through an assessment of international activities in the field. The methodology incorporated bibliographic research, examination of relevant literature, and

[1]Oden, J.T., Belytschko, T., Fish, J., Hughes, T.J.R., Johnson, C., Keyes, D., Laub, A., Petzold, L., Srolovitz, D. and Yip, A. (2006). *Simulation-based Engineering Science: Revolutionizing Engineering Science Through Simulation* (National Science Foundation, Arlington, VA), http://www.nsf.gov/pubs/reports/sbes_final_report.pdf.

in particular, site visits and personal contacts with leaders in the field around the world. Through these means, the intent was to identify specific technical strengths and weaknesses; sites where the most advanced activities are occurring; what research challenges and apparent roadblocks exist; and what opportunities are just appearing.

One important change was that "SBES" was broadened to "SBE&S" to represent its multidisciplinary nature more effectively. By considering SBE&S as the overarching topic, the study also explored impacts that computer simulation is having on pure, curiosity-driven science, on problem-driven science, and on engineering development, design, prediction, decision-making, and manufacturing.

These goals resonated within NSF and with other agencies. NSF's Engineering Directorate was the lead sponsor of this study, joined by programs from NSF's Mathematics and Physical Sciences Directorate, the Department of Defense, the Department of Energy, the National Aeronautics and Space Administration, the National Institute for Biomedical Imaging and Bioengineering, the National Library of Medicine, and the National Institute of Standards and Technology. We selected the World Technology Evaluation Center, Inc. (WTEC) to arrange bibliometrics, trip logistics, and diverse aspects of aiding the study and its report preparation.

We are grateful for all these commitments and for the commitment and diligence of the study participants. They include the study panel, the advisory panel, and the gracious hosts of our site visits, who all contributed thoughtful, informative analysis and discussion. Our goal is now to use these insights to enhance and guide SBE&S activities throughout the nation.

P. R. Westmoreland, K. P. Chong, and C. V. Cooper
National Science Foundation
January 2009

ABSTRACT

This WTEC panel report assesses the international research and development activities in the field of Simulation-Based Engineering and Science (SBE&S). SBE&S involves the use of computer modeling and simulation to solve mathematical formulations of physical models of engineered and natural systems. SBE&S today has reached a level of predictive capability that it now firmly complements the traditional pillars of theory and experimentation/observation. As a result, computer simulation is more pervasive today – and having more impact – than at any other time in human history. Many critical technologies, including those to develop new energy sources and to shift the cost-benefit factors in healthcare, are on the horizon that cannot be understood, developed, or utilized without simulation. A panel of experts reviewed and assessed the state of the art in SBE&S as well as levels of activity overseas in the broad thematic areas of life sciences and medicine, materials, and energy and sustainability; and in the crosscutting issues of next generation hardware and algorithms; software development; engineering simulations; validation, verification, and uncertainty quantification; multiscale modeling and simulation; and SBE&S education. The panel hosted a U.S. baseline workshop, conducted a bibliometric analysis, consulted numerous experts and reports, and visited 59 institutions and companies throughout East Asia and Western Europe to explore the active research projects in those institutions, the computational infrastructure used for the projects, the funding schemes that enable the research, the collaborative interactions among universities, national laboratories, and corporate research centers, and workforce needs and development for SBE&S.

The panel found that SBE&S activities abroad are strong, and compete with or lead the United States in some strategic areas. Both here and abroad, SBE&S is changing the way disease is treated, the way surgery is performed and patients are rehabilitated, and the way we understand the brain; changing the way materials and components are designed, developed, and used in all industrial sectors; and aiding in the recovery of untapped oil, the discovery and utilization of new energy sources, and the way we design sustainable infrastructures. Data-intensive and data-driven applications were evident in many countries. Achieving millisecond timescales with molecular resolution for proteins and other complex matter is now within reach due to new architectures and algorithms. The fidelity of engineering simulations is being improved through inclusion of physics and chemistry. There is excitement about the opportunities that petascale computers will afford, but concern about the ability to program them. Because fast computers are now so affordable, and several countries are committed to petascale computing and beyond, what will distinguish us from the rest of the world is our ability to do SBE&S better and to exploit new architectures we develop before those architectures become ubiquitous. Inadequate education and training of the next generation of computational scientists and engineers threatens global as well as U.S. growth of SBE&S. A persistent pattern of subcritical funding overall for SBE&S threatens U.S. leadership and continued needed advances, while a surge of strategic investments in SBE&S abroad reflects recognition by those countries of the role of simulation in advancing national competitiveness and its effectiveness as a mechanism for economic stimulus.

There are immediate opportunities to strengthen the U.S. capability for SBE&S through strategic investments in industry-driven partnerships with universities and national laboratories; new and sustained mechanisms for supporting R&D in SBE&S; and a new, modern approach to educating and training the next generation of researchers in high performance computing, modeling and simulation for scientific discovery and engineering innovation. Key findings in the three thematic domain areas of the study and in the crosscutting areas and technologies that support SBE&S reinforce these overarching findings in specific ways.

ADDENDUM

The importance of Simulation-Based Engineering and Science as a critical enabler in technological and economic development has been increasingly recognized worldwide since the initial release of this study in early 2009. Indeed, the role of SBE&S and the related fields of information technology and high performance computing in national competitiveness is reflected in current US planning and budget documents, and is a critical part of the current administration's science and technology vision. The race to maintain US leadership in SBE&S has never been more heated, and the landscape is changing rapidly. In 2010, China emerged as a frontrunner in the supercomputer race with a machine capable of performing calculations at a speed of 2.5 petaflops, using U.S.-made chips and proprietary Chinese interconnect technology. The next generation of Chinese supercomputer is expected to be built entirely on Chinese technology, clearly elevating China to the level of technology superpower in the field of HPC. The detailed findings of the present study with respect to past and growing commitment to SBE&S research, education, and training in various countries, including China, are especially relevant given these technological advances. The findings provide both a unique look at China's path toward their present capabilities in HPC and simulation-based scientific discovery and engineering innovation, and also their future ability to compete with nations around the world, including the U.S., in energy, materials, life sciences, and other areas of great strategic importance.

Sharon C. Glotzer
April 5, 2011

EXECUTIVE SUMMARY

1 Background

Simulation-Based Engineering and Science (SBE&S) involves the use of computer modeling and simulation to solve mathematical formulations of physical models of engineered and natural systems. Today we are at a "tipping point" in computer simulation for engineering and science. Computer simulation is more pervasive today — and having more impact — than at any other time in human history. No field of science or engineering exists that has not been advanced by, and in some cases transformed by, computer simulation. Simulation has today reached a level of predictive capability that it now firmly complements the traditional pillars of theory and experimentation/observation. Many critical technologies are on the horizon that cannot be understood, developed, or utilized without simulation. At the same time, computers are now affordable and accessible to researchers in every country around the world. The near-zero entry-level cost to perform a computer simulation means that anyone can practice SBE&S, and from anywhere. Indeed, the world of computer simulation is becoming flatter every day. At the same time, U.S. and Japanese companies are building the next generation of computer architectures, with the promise of thousand-fold or more increases of computer power coming in the next half-decade. These new massively manycore computer chip architectures will allow unprecedented accuracy and resolution, as well as the ability to solve the highly complex problems that face society today. Problems ranging from finding alternative energy sources to global warming to sustainable infrastructures, to curing disease and personalizing medicine, are *big* problems. They are complex and messy, and their solution requires a partnership among experiment, theory, and simulation working across all of the disciplines of science and engineering. There is abundant evidence and numerous reports documenting that our nation is at risk of losing its competitive edge. Our continued capability as a nation to lead in simulation-based discovery and innovation is key to our ability to compete in the twenty-first century.

To provide program managers in U.S. research agencies and decision makers with a better understanding of the status, trends and levels of activity in SBE&S research abroad, these agencies sponsored the WTEC International Assessment of R&D in Simulation-Based Engineering and Science. Sponsors included

- National Science Foundation;
- Department of Energy;
- National Institutes of Health (NIBIB, National Library of Medicine);
- NASA;
- National Institute of Standards and Technology;
- Department of Defense.

The study was designed to gather information on the worldwide status and trends in SBE&S research and to disseminate this information to government decision makers and the research community. A panel of experts reviewed and assessed the state of the art as well as levels of activity overseas in the broad thematic areas of SBE&S in life sciences and medicine, materials, in energy and sustainability; and in the crosscutting issues of next generation hardware and algorithms; software development; engineering simulations; validation, verification, and uncertainty quantification; multiscale modeling and simulation; and education. To provide a basis for comparison, the panel hosted a U.S. Baseline workshop at the National Science Foundation (NSF) on November 1–2, 2007. Following the workshop, a WTEC panel of U.S. experts visited 59 sites in Europe and Asia involved in SBE&S research. Information gathered at the site visits, along with individual conversations with experts, access to reports and research publications, and a bibliometric analysis, provided the basis for the assessment. An advisory panel provided additional feedback to the study panel.

In this executive summary, we first present the panel's overarching findings concerning major trends in SBE&S research and development, threats to U.S. leadership in SBE&S, and opportunities for gaining or reinforcing the U.S. lead through strategic investments. We then highlight the most notable findings in each of the three thematic areas and each of the crosscutting areas of the study. Throughout this summary, key findings are shown in bold italics, with additional text provided to support and amplify these findings.

2 Major Trends in SBE&S Research and Development

Within the thematic areas of the study, *the panel observed that SBE&S is changing the way disease is treated and surgery is performed, the way patients are rehabilitated, and the way we understand the brain; the way materials and components are designed, developed, and used in all industrial sectors; and aiding in the recovery of untapped oil, the discovery and utilization of new energy sources, and the way we design sustainable infrastructures.* In all of these areas, there were ample examples of critical breakthroughs that will be possible in the next decade through application of SBE&S, and in particular through the following four major trends in SBE&S:

Finding 1: Data-intensive applications, including integration of (real-time) experimental and observational data with modeling and simulation to expedite discovery and engineering solutions, were evident in many countries, particularly Switzerland and Japan.

Modeling and simulation of very large data sets characterized some of the most cutting-edge examples of SBE&S in life sciences and medicine, especially with regards to systems biology. Big data already plays a prominent role in elementary particle physics, climate modeling, genomics, and earthquake prediction. The trend will become even more prominent globally with petascale computing.

Finding 2: Achieving millisecond time scales with molecular resolution for proteins and other complex matter is now within reach using graphics processors, multicore CPUs, and new algorithms.

Because this is the time scale on which many fundamental biomolecular processes occur, such as protein folding, this represents an enormous breakthrough and many opportunities for furthering our understanding of processes important for biomedical and life science applications. These same speed-ups will further revolutionize materials simulation and prediction, allowing the level of resolution needed for materials design as well as modeling chemical and physical processes such as those relevant for biofuels, batteries, and solar cells.

Finding 3: The panel noted a new and robust trend towards increasing the fidelity of engineering simulations through inclusion of physics and chemistry.

Although the panel did not see much evidence of sophisticated approaches towards validation and verification, uncertainty quantification, or risk assessment in many of the simulation efforts, which would complement the use of physics-based models in engineering simulations, there was widespread agreement that this represents a critical frontier area for development in which the United States currently has only a slight lead.

Finding 4: The panel sensed excitement about the opportunities that petascale speeds and data capabilities would afford.

Such capabilities would enable not just faster time to solution, but also the ability to tackle important problems orders-of-magnitude more complex than those that can be investigated today with the needed level of predictiveness. Already, inexpensive graphics processors today provide up to thousand-fold increases in speed on hundreds of applications ranging from fluid dynamics to medical imaging over simulations being performed just a year or two ago.

3 Threats to U.S. Leadership in SBE&S

The panel identified three overarching issues that specifically threaten U.S. leadership in SBE&S.

Finding 1: The world of computing is flat, and anyone can do it. What will distinguish us from the rest of the world is our ability to do it better and to exploit new architectures we develop before those architectures become ubiquitous?

First and foremost, many countries now have and use HPC — *the world is flat!* A quick look at the most recent Top 500 list of supercomputers shows that Japan, France, and Germany in particular have world-class resources. They also have world class faculty and students and are committed to HPC and SBE&S for the long haul. In terms of hardware, Japan has an industry–university–government roadmap out to 2025 (exascale), and Germany is investing nearly US$1 billion in a new push towards next-generation hardware in partnership with the European Union. It is widely recognized that it is relatively inexpensive to start up a new SBE&S effort, and this is of particular importance in rapidly growing economies (e.g., India, China, Finland). Furthermore, already there are more than 100 million NVIDIA graphics processing units with CUDA compilers distributed worldwide in desktops and laptops, with potential code speedups of up to a thousand-fold in virtually every sector to whomever rewrites their codes to take advantage of these new general programmable GPUs.

Aggressive, well-funded initiatives in the European Union may undermine U.S. leadership in the development of computer architectures and applied algorithms. Examples of these initiatives include the Partnership for Advanced Computing in Europe (PRACE) which is a coalition of 15 countries and led by Germany and France, and based on the ESFRI Roadmap (ESFRI 2006); TALOS — Industry–government alliance to accelerate HPC solutions for large-scale computing systems in Europe; and DEISA — Consortium of 11 leading European Union national supercomputing centers to form the equivalent of the U.S. TeraGrid. There is also some flux, with some alliances dissolving and new consortia being formed. Already, the European Union leads the United States in theoretical algorithm development, and has for some time; these new initiatives may further widen that lead and create new imbalances.

Finding 2: Inadequate education and training of the next generation of computational scientists threatens global as well as U.S. growth of SBE&S. This is particularly urgent for the United States; unless we prepare researchers to develop and use the next generation of algorithms and computer architectures, we will not be able to exploit their game-changing capabilities.

There was grave concern, universally voiced at every site in every country including the United States, that today's computational science and engineering students are ill-prepared to create and innovate the next generation of codes and algorithms needed to leverage these new architectures. Much of this appears to arise from insufficient exposure to computational science and engineering and underlying core subjects beginning in high school and undergraduate and continuing through graduate education and beyond. Increased topical specialization beginning with graduate school was noted as a serious barrier to a deep foundation in simulation and supporting subjects. There is a clear gap in the preparation students receive in high performance computing and what they need to know to develop codes for massively parallel computers, let alone petascale and massively multicore architectures. Worldwide, students are not learning to "program for performance." Nearly universally, the panel found concern that students use codes primarily as black boxes, with only a very small fraction of students learning proper algorithm and software development, in particular with an eye towards open-source or community code development. Students receive no real training in software engineering for sustainable codes, and little training if any in uncertainty quantification, validation and verification, risk assessment or decision making, which is critical for multiscale simulations that bridge the gap from atoms to enterprise. Moreover, in many countries, the very best computational science and engineering students leave research for finance companies (this was, until recently, a particularly huge problem in Switzerland). Despite the excitement about manycore and petascale computing, the panel noted universal concern that the community is not prepared to take advantage of these hardware breakthroughs because the current generation of algorithms and software must be rethought in the context of radically new architectures that few know how to program.

Finding 3: A persistent pattern of subcritical funding overall for SBE&S threatens U.S. leadership and continued needed advances amidst a recent surge of strategic investments in SBE&S abroad that reflects recognition by those countries of the role of simulations in advancing national competitiveness and its effectiveness as a mechanism for economic stimulus.

It is difficult to draw a precise boundary around funding specific to SBE&S whether globally or within the United States due to its interdisciplinary nature. Nevertheless, we can consider the total funding for the entire information technology/computer science/computational science/cyberinfrastructure landscape (NITRD in the case of the United States) and identify within this large space specific, high-profile programs in which simulations play a central, enabling role (e.g., SCIDAC, ITR, Cyber-enabled Discovery, etc.). This exercise allows an "apples to apples" comparison with activities around the world, such as the notable growth spurts in SBE&S funding in China and Germany.

As noted in Oden *et al.* (2006), the pattern of subcritical funding for SBE&S is masked in the overall budget across the landscape for information and computational technology ($3.5 billion NITRD FY2009 request) (OMB, 2008). The situation is particularly dire in light of the role of SBE&S in achievement of national priorities in energy, health, and economic development and alarming when viewed in comparison with the panel's findings of investments targeting SBE&S across the globe. The panel found that the level of SBE&S investments abroad reflect the recognition of the strategic role of simulations in advancing national competitiveness as well as a tacit recognition of their effectiveness as an economic stimulus mechanism. Consequently, funding for SBE&S is surging in both peer OECD partners (e.g., Germany and Japan) as well as in rapidly developing economies (e.g., China). Evidence for this is exemplified by the following observations:

- In Germany, specific and focused investments in SBE&S are patterned along the recommendations in the 2006 NSF blue ribbon panel report on SBES (Oden *et al.*, 2006) as part of the 20+% year-on-year increase in funding for research. As a consequence of this new funding, Germany already exhibits many of the innovative organizational and collaborative structures deemed to be the most promising for advancing SBE&S in the context of energy, medicine, and materials research. The panel observed extensive restructuring of universities to enable more interdisciplinary work and strong university–industry partnerships.
  - One example of such a partnership is between the Fraunhofer IWM and Karlsruhe University. The IWM has an annual budget of \$16 million/yr, with 44% coming from industry, and 50% supporting research in SBE&S.
- Funding for SBE&S in China is robust and an integral part of the national strategy for upgrading the R&D infrastructure of the top and mid-tier universities. Most recently, it has become apparent that China's SBE&S funding within the context of a \$billion-plus increase in funding for universities, is an integral element of their economic stimulation strategy.
  - Although China is not yet a strong U.S. competitor in SBE&S, their "footprint" is changing rapidly. China contributes 13% of the world's output in simulation papers, second to the United States at 26% (from 2003 to 2005) and third to the 12 leading Western European countries combined (32%) and growing fast (although they publish in less than first-tier journals and the papers are generally cited less often). China also had the highest relative commitment to simulation research over the decade 1996–2005, with 50% more papers than expected on the basis of its overall output (the US published about 10% fewer papers than expected on this basis). The panel found nonuniform quality overall, but saw many high quality examples on par with the best examples anywhere. There was a palpable sense of a strategic change in the direction of funding and research towards innovation, combined with a clearly articulated *recognition by industry and the Chinese government that innovation*

requires simulation. China's S&T budget has doubled every 5 years since 1990, with 70% of the funding going to the top 100 universities, which produce 80% of all PhDs, support 70% all graduates, 50% of all international students, and 30% of all undergraduates. There is a specific recognition by the Vice Minister of Education of the need to train a new generation of "computationally–savvy" students with solid training in analytical thinking, and substantial new funds to do this in programs under his control. One example is the 211 Fund, which is the equivalent of roughly US$1 billion/year; all new projects must have an integrated simulation component.

- For Japan, in addition to the expected level of funding for the successor to the Earth Simulator (Life Simulator), the "accomplishment-based funding" for systems biology including SBE&S activities was a noteworthy finding of the SBE&S funding landscape in Japan. Japan leads the US in bridging physical systems modeling to social-scale engineered systems, in particular by the introduction of large-scale agent-based simulations.
- The United Kingdom has had a decade-plus experience with community-based code development for SBE&S in the form of the highly successful Collaborative Computational Projects (CCP). Although the program is now limited in scope to maintenance of codes, its multiple contributions to the standard tool set of the global SBE&S infrastructure far exceed the scale of the investment, suggesting that this mechanism merits further consideration both by the United States and the global community.
- Although outside the scope of the panel's trips to Europe, China, and Japan, recent investments in R&D and in SBE&S by both Singapore and Saudi Arabia are particularly noteworthy. With substantial involvement by U.S. faculty and enormous financial investment, Singapore has reinvented their R&D infrastructure and is carrying out world-class research in many strategic areas, including new programs in computational science and engineering. Stunning investments by Saudi Arabia in the King Abdullah University of Science and Technology (KAUST) (with $20B endowment) has generated substantial interest around the world, with some of the best faculty from the United States and Europe leading many of their major activities. A multimillion dollar IBM Blue Gene P supported by IBM provides access to supercomputing resources for KAUST researchers and students. In particular, there is substantial concern that KAUST, with its state-of-the-art facilities and

generous graduate scholarships, will be able to attract away the best and brightest Asian students, who have for many years come to the United States for graduate study and remained to become leaders in high-tech industries. It is particularly interesting to note that Applied Mathematics and Computational Science, the core of SBE&S, is one of KAUST's very first graduate programs.
- The panel found healthy levels of SBE&S funding for internal company projects in our visits to the industrial R&D centers, underscoring industry's recognition of the cost-effectiveness and timeliness of SBE&S research. The mismatch *vis a vis* the public-sector's investment in SBE&S hinders workforce development. Notably, the panel saw many examples of companies (including U.S. auto and chemical companies) working with E.U. groups rather than U.S. groups for "better IP agreements."
- Because SBE&S is often viewed within the United States more as an enabling technology for other disciplines, rather than a discipline in its own right, investment in and support of SBE&S is often not prioritized as it should be at all levels of the R&D enterprise. As demonstrated in the 2005 PITAC report (Benioff, 2005), the panel found that investment in computational science in the United States and the preparation of the next generation of computational researchers remains insufficient to fully leverage the power of computation for solving the biggest problems that face us, as a nation going forward.

4 Opportunities for the United States to Gain or Reinforce Lead in SBE&S through Strategic Research and Investments

The panel identified several specific ways in which the United States could gain or reinforce its lead in SBE&S, and these are described in detail in the report. Three of the most important findings in terms of opportunities are:

Finding 1: There are clear and urgent opportunities for industry-driven partnerships with universities and national laboratories to hardwire scientific discovery and engineering innovation through SBE&S. This would lead to new and better products, as well as development savings both financially and in terms of time. This idea is exemplified in a new National Academies' report on Integrated Computational Materials Engineering (ICME) (NRC, 2008), which found a reduction in development time from 10–20 yrs to 2–3 yrs with a concomitant return on investment of 3:1 to 9:1.

Finding 2: There is a clear and urgent need for new mechanisms for supporting R&D in SBE&S. Particular attention must be paid to the way that we support and reward the long-term development of algorithms, middleware, software, code maintenance and interoperability.

For example, the panel found that *community code development projects are much stronger within the European Union than the United States, with national strategies and long-term support.* Many times the panel was told that the United States is an "unreliable partner" in these efforts due to our inability to commit for longer than typically three years at a time. Both this perception and the reality means that the United States has little influence over the direction of these community codes and at the same time is not developing large codes of its own. As one alarming example of this, six of the top seven electronic structure codes for material physics today come from the European Union, despite the fact that the original science codes are from the United States (and for which the United States won the Nobel Prize), and the scientific advances underlying the codes are distributed evenly between the United States and European Union. Indeed, the U.S. lead in scientific software applications has decreased steadily since 1998. One consequence of this lack of involvement and influence is that, in some cases, codes may not be available to U.S. researchers. As one critical example, a major new code for electronic transport from Spain/United Kingdom is effectively off limits to U.S. defense labs.

The panel found universal agreement (although the disparity is greatest in the United States) that *investment in algorithm, middleware, and software development lags behind investment in hardware, preventing us from fully exploiting and leveraging new and even current architectures. This disparity threatens critical growth in SBE&S capabilities needed to solve important worldwide problems as well as many problems of particular importance to the U.S. economy and national security.*

A related issue is evidenced lack of support and reward for code development and maintenance. It is widely recognized within the SBE&S community that the timescale to develop large complex code often exceeds the lifetime of a particular generation of hardware. Moreover, although great scientific advances achieved through the use of a large complex code is highly lauded, the development of the code itself often goes unrewarded. The United Kingdom, which once led in supporting long-term scientific software development efforts, does not provide the support it once did.

The panel found worldwide recognition that progress in SBE&S requires crossing disciplinary boundaries, which is difficult in many countries due to outdated university structures. The United States is perceived to be a leader in pulling together interdisciplinary teams for large, complex problems, with the caveats noted above.

Finding 3: There is a clear and urgent need for a new, modern approach to educating and training the next generation of researchers in high performance computing specifically, and in modeling and simulation generally, for scientific discovery and engineering innovation.

Particular attention must be paid to teaching fundamentals, tools, programming for performance, verification and validation, uncertainty quantification, risk analysis and decision making, and programming the next generation of massively multicore architectures. At the same time, students must gain deep knowledge of their core discipline.

5 Key Study Findings

Here, we briefly summarize the panel's main findings in each of the three primary thematic domains and the crosscutting areas. Each of these findings as well as additional observations are discussed in detail in the succeeding chapters of this report. Many of the findings underpinning individual crosscutting areas echo those of the thematic domains, giving rise to occasional but telling redundancies that serve to amplify the importance of many crosscutting technologies and issues to the application of simulation to particular science and engineering domains.

5.1 *Thematic Area: Life Sciences and Medicine*

There is an unprecedented opportunity in the next decade to make game-changing advances in the life sciences and medicine through simulation.

Finding 1: Predictive biosimulation is here.

New techniques for modeling variability in patient physiology, as well as new software platforms, lower-cost computational resources, and an explosion of available data, have fueled recent advances in "predictive biosimulation" — the *dynamic simulation of biological systems*. This is a major advance in the use of SBE&S in the medical field, moving beyond simple chemistry and molecular structure. Predictive biosimulation has been used to support the development of new therapeutic drugs, meeting the pharmaceutical industry's need for new approaches to deal with the complexity of disease and the increased risk and expense of R&D.

- Several factors are likely to accelerate adoption of biosimulation technology in the pharmaceutical industry. The FDA, through its Critical Path Initiative, has recognized the importance of modeling and simulation and is committing more resources toward making biosimulation part of the evaluation process for new therapies. Academic groups and companies are increasingly specializing in the field and building active collaborations with industry. Pharmaceutical companies are creating and expanding internal systems biology groups that incorporate biological modeling.
- Modeling of whole organs and complete physiological systems is becoming possible.
  - Modeling of complete cardiovascular systems provides information and insight both for medical research and for the evaluation of surgical options.
  - Modeling provides information both for cancer research and for laser surgery.
  - Neuromuscular biomechanics seeks fundamental understanding of the mechanisms involved in the production of movement and is motivated by opportunities to improve treatments for cerebral palsy.
  - Biomechanical simulation is used in surgical planning and design of physical therapies.
  - Rapidly increasing amounts of data, new algorithms, and fast computers make modeling the brain in full experimental detail, a realistic goal for the foreseeable future.

Finding 2: Pan-SBE&S synergy argues for a focused investment of SBE&S as a discipline.

The exciting possibilities of predictive biosimulation depend on a collective set of advances in SBE&S capabilities, a finding in common with the other two themes. This dependence buttresses the notion of a focused advancement of SBE&S capabilities as a discipline. For predictive biosimulation to realize its potential impacts in life sciences and medicine will require:

- Large, focused multidisciplinary teams. Much of the research in this area requires iteration between modeling and experiment.

- Integrated, community-wide software infrastructure for dealing with massive amounts of data, and addressing issues of data provenance, heterogeneous data, analysis of data, and network inference from data.
- High-performance algorithms for multiscale simulation on a very large range of scales and complexities.
- High-performance computing and scalable algorithms for multicore architectures. In particular, petascale computing can finally enable molecular dynamics simulation of macromolecules on the biologically interesting millisecond timescale.
- Techniques for sensitivity and robustness analysis, uncertainty analysis, and model (in)validation.
- Visualization techniques, applicable to massive amounts of data, that can illustrate and uncover relationships among data.
- Appropriately trained students who are conversant in both the life sciences and SBE&S and can work in multidisciplinary teams.

Finding 3: Worldwide SBE&S capabilities in life sciences and medicine are threatened by lack of sustained investment and loss of human resources.

World-class research exists in SBE&S in molecular dynamics, systems biology, and biophysical modeling throughout the United States, Europe, and Japan, and, in some research areas, of China. The United States has historically played a leadership role in a number of research areas that have driven the development of new algorithms, architectures, and applications for molecular dynamics (MD) simulations. However, this role is now endangered by a lack of sustained investment at all federal agencies, and by the loss of PhD students and faculty to other disciplines and to the private sector.

- The quality of leading researchers within this field in the United States is comparable to that of leading researchers in Europe and Japan.
- Infrastructure (access to computing resources and software professionals) and funding models to support ambitious, visionary, long-term research projects are much better in Europe and Japan than in the United States.
- Funding models and infrastructure to support multi-investigator collaborations between academia and industry are much more developed in Europe than in the United States.
- Support for the development of community software for biosimulation is much stronger in Europe and Japan than in the United States in recent years, following the decline in DARPA funding in this area.

5.2 *Thematic Area: Materials*

SBE&S has long played a critical role in materials simulation. As experimental techniques for synthesizing and characterizing new materials become increasingly sophisticated, the demand for theoretical understanding and design guidelines is driving a persistently increasing need for materials simulation.

Finding 1: Computational materials science and engineering is changing how new materials are discovered, developed, and applied, from the macroscale to the nanoscale.

- This is particularly so in nanoscience, where it is now possible to synthesize a practically infinite variety of nanostructured materials and combine them into new devices or systems with complex nanoscale interfaces. The experimental tools of nanoscience — such as scanning probes, neutron scattering, and various electron microscopies — all require modeling to understand what is being measured. Hence, the demand for high-fidelity materials simulation is escalating rapidly, and the tools for verification of such simulations are becoming increasingly available. The nanoscale is particularly ripe for SBE&S since both experiment and simulation can access the same length and time scales.

Finding 2: World-class research in all areas of materials simulation is to be found in the United States, Europe, and Asia; the United States leads in some, but not all, of the most strategic of these.

- Algorithm innovation takes place primarily in the United States and Europe, some in Japan, with little activity to date in this area in China.
- There is a rapid ramping-up of materials simulation activities in some countries, particularly China and Germany.
  - Some of these efforts in Germany involve close partnerships between industry, government, and academia (e.g., through the Fraunhofer Institutes). New efforts are sprouting against a backdrop of an extraordinary revival of science in Germany, tied to recent increases in the Deutsche Forschungsgemeinschaft (DFG, German Research Foundation, the equivalent of the U.S. National Science Foundation) budget. In the most recent years, DFG funding jumped over 20%, following on three years of increases in the 5–10% range.

— Historically China has been mostly concerned with applications; however, as part of the Chinese government's current emphasis on supporting creativity and innovation in both the arts and sciences, one can anticipate increased activity in this area in China.

Finding 3: The United States' ability to innovate and develop the most advanced materials simulation codes and tools in strategic areas has been eroding and continues to erode.

- The United States is at an increasingly strategic disadvantage with respect to crucial, foundational codes, as it has become increasingly reliant on codes developed by foreign research groups. Political considerations can make those codes effectively unavailable to U.S. researchers, particularly those in Department of Defense laboratories.
- Many large codes, both open source and non-open source, require collaboration among large groups of domain scientists, applied mathematicians, and computational scientists. However, there is much greater collaboration among groups in materials code development in Europe compared to the United States. There appear to be several reasons for this:
  — The U.S. tenure process and academic rewards systems suppress collaboration.
  — Funding, promotion, and awards favor high-impact science (publications in Nature and Science, for example), while the development of simulation tools is not considered to be high-impact science. Yet, these tools (which can take many years to develop) are often the key factor in enabling the high-impact science.
- The utility of materials simulation codes for practical application is threatened by the lack of standards for interoperability of codes. The CAPE-OPEN effort undertaken in the chemical process simulation (computer-aided process engineering, or CAPE) field to develop interoperability standards in that field (http://www.colan.org) is an example of an approach that could benefit the materials community.
- The current practice of short-term (less than five years, typically three-year cycles) funding is insufficient for the United States to develop key materials simulation codes that will run on the largest U.S. computing platforms and on future many-core processors. Such codes will be needed to explore the next generation of materials and to interpret experiments on the next generation of scientific instruments. Longer term funding (minimum five years) of multidisciplinary teams is needed.

- Training is a critical issue worldwide, as undergraduate students in the physical and chemical sciences and engineering domains that feed into the computational materials discipline are increasingly illiterate in computer programming. The situation is perhaps most acute in the United States.

5.3 *Thematic Area: Energy and Sustainability*

The search for alternative energy sources and cleaner fuels, and for long-term sustainable infrastructures, is one of the greatest and most serious challenges facing the planet over the next decade. The role for SBE&S is enormous and urgent. Broad and extensive simulation-based activities in energy and sustainability exist in all regions. However, the scarcity of appropriately trained students in the U.S. (relative to Europe and Asia) is seen as a substantial bottleneck to progress.

Finding 1: In the area of transportation fuels, SBE&S is critical to stretch the supply and find other sources.

- Recognition that future energy sources for transportation cannot be substituted for traditional sources in the short term is motivating the simulation of oil reservoirs to optimize recovery, refinery processes to optimize output and efficiency, and the design of more fuel-efficient engines, including diesel and hybrid engines.
- The use of SBE&S in oil production and optimizing the fossil-fuel supply chain has traditionally been led by the United States. The panel found significant activity in Europe, including new funding models and leveraging strengths in development of community codes for applications to this important problem.

Finding 2: In the discovery and innovation of alternative energy sources — including biofuels, batteries, solar, wind, nuclear — SBE&S is critical for the discovery and design of new materials and processes.

- Nuclear energy is a major source for electricity in the United States and abroad. Rapid progress and increased investment in SBE&S-driven materials design for nuclear waste containment are urgently needed. France leads the world in nuclear energy and related research (e.g., containment).

- Tax incentives for green sources of energy, despite lower energy density, are driving extensive R&D efforts for more efficient wind energy and tidal energy systems, especially in Europe.
- Alternative fuels such as biodiesel are constituted from a complex mixture of plant-derived organic chemicals. Optimal engine combustion designs require the thermophysical properties of such complex mixtures, and in view of the myriad possible compositions faced by the experimentalist, simulations provide the only viable option.

Finding 3: Petascale computing will allow unprecedented breakthroughs in sustainability and the simulation of ultra-large-scale sustainable systems, from ecosystems to power grids to whole societies.

- Simulations to evaluate the performance of large-scale and spatially distributed systems such as power transmission systems that are subject to highly uncertain natural hazards (earthquake, hurricanes, flood) hold important promise for the design of sustainable infrastructures and for disaster prediction and prevention.
- Japan is pushing forward the modeling of human behavior through massive agent-based simulations enabled by their Life Simulator, the successor to Japan's Earth simulator.
- It is well recognized that even with technological breakthroughs in viable alternative sources of energy, the lack of ease of substitution due to a lack of critical infrastructure (e.g., diesel stations for diesel engines) and public acceptance is a major challenge. The modeling and simulation of integration of alternative sources into the power grid is a growing area, revealing technological issues as well as how the public adopts the alternative sources, behaves, etc.
- Simulation-based engineering is being used to create "smart" infrastructures.
- The United States leads in modeling of large-scale infrastructure systems, but the gap is closing, particularly in Asia.

5.4 *Crosscutting Issues: Next Generation Architectures and Algorithms*

Finding 1: The many orders-of-magnitude in speedup required to make significant progress in many disciplines will come from a combination of synergistic advances in hardware, algorithms, and

software, and thus investment and progress in one will not pay off without concomitant investments in the other two.

In the last decade, large investments have been made for developing and installing new computer architectures that are now breaking the petaflop barrier, and the exascale era will be here within 10 years. However, recent investments in mathematical algorithms and high-quality mathematical software are lagging behind. Specifically, on the algorithmic front, advances in linear solvers, high-order spatio-temporal discretization, domain decomposition, and adaptivity are required to match the new manycore and special-purpose computer architectures and increase the "effective" performance of these expensive supercomputers.

Finding 2: The United States leads both in computer architectures (multicores, special-purpose processors, interconnects) and applied algorithms (e.g., ScaLAPACK, PETSC), but aggressive new initiatives around the world may undermine this position. Already, the European Union leads the United States in theoretical algorithm development, and has for some time.

In particular, the activities of the Department of Energy laboratories have helped to maintain this historic U.S. lead in both hardware and software. However, the picture is not as clear on the development of *theoretical* algorithms; several research groups across Europe (e.g., in Germany, France, and Switzerland) are capitalizing on new "priority programs" and are leading the way in developing fundamentally new efforts for problems on high dimensions, on O(N) algorithms, and specifically on fast linear solvers related to hierarchical matrices, similar to the pioneering work in Europe on multigrid in the 1980s. Initiatives such as the "Blue Brain" at EPFL and "Deep Computing" at IBM-Zurich (see site reports in Appendix C) have set ambitious goals far beyond what the U.S. computational science community can target at present. It is clear that barriers to progress in developing next-generation architectures and algorithms are increasingly on the theoretical and software side and that the return on investment is more favorable on the theoretical and software side. This is manifested by the fact that the "half-time" of hardware is measured in years, whereas the "half-time" of software is measured in decades and theoretical algorithms transcend time.

Finding 3: Although the United States currently leads in the development of next-generation supercomputers, both Japan and Germany have been, and remain, committed over the

long-term to building or acquiring leadership-class machines and capability, and China is now beginning to develop world-class supercomputing infrastructure.

5.5 *Crosscutting Issues: Scientific and Engineering Simulation Software Development*

Modern software for simulation-based engineering and science is sophisticated, tightly coupled to research in simulation models and algorithms, and frequently runs to millions of lines of source code. As a result, the lifespan of a successful program is usually measured in decades, and far surpasses the lifetime of a typical funding initiative or a typical generation of computer hardware.

Finding 1: Around the world, SBE&S relies on leading edge (supercomputer class) software used for the most challenging HPC applications, mid-range and desktop computing used by most scientists and engineers, and everything in between.

Simulation software is too rich and too diverse to suit a single paradigm. Choices as disparate as software targeting cost-effective computer clusters versus leading edge supercomputers, or public domain software versus commercial software, choices of development tools, etc., are mapped across the vast matrix of application disciplines. The best outcomes seem to arise from encouraging viable alternatives to competitively co-exist, because progress driven by innovation occurs in a bottom-up fashion. Thus strategic investments should balance the value of supporting the leading edge (supercomputer class) applications against the trailing vortex (mid-range computing used by most engineers and scientists).

Finding 2: Software development leadership in many SBE&S disciplines remains largely in U.S. hands, but in an increasing number of areas it has passed to foreign rivals, with Europe being particularly resurgent in software for mid-range computing, and Japan particularly strong on high-end supercomputer applications. In some cases, this leaves the United States without access to critical scientific software.

In many domains of SBE&S, the U.S. relies on codes developed outside the U.S. despite having led previously in the theoretical research that formed the intellectual basis for such codes. A major example of that is in first principles codes for materials, the theoretical framework for which the United States won the 1998 Nobel Prize in Chemistry. Just one decade later, the WTEC panel learned of American defense researchers denied access to foreign-developed electronic transport software as a matter of principle, with no equivalent U.S. codes available as a substitute.

Finding 3: The greatest threats to U.S. leadership in SBE&S come from the lack of reward, recognition and support concomitant with the long development times and modest numbers of publications that go hand-in-hand with software development; the steady erosion of support for first-rate, excellence-based single-investigator or small-group research in the United States; and the inadequate training of today's computational science and engineering students — the would-be scientific software developers of tomorrow.

Fundamentally, the health of simulation software development is inseparable from the health of the applications discipline it is associated with. Therefore, the principal threat to U.S. leadership comes from the steady erosion of sustained support for first-rate, excellence-based single-investigator or small-group research in the United States. A secondary effect that is specific to software development is the distorting effect that long development times and modest numbers of publications have on grant success rates. Within applications disciplines, it is important to recognize and reward the value of software development appropriately, in balance with the direct exploration of phenomena. Software development benefits industry and society through providing useful tools too expensive, long-term and risky to be done as industrial R&D, trains future scientists and engineers to be builders and not just consumers of tools, and helps to advance the simulation state of the art. Future investments in software development at the applications level are best accomplished as part of re-invigorating American physical and biological science generally. More specific investments in simulation software can be justified on the basis of seeking targeted leadership in areas of particular technological significance, as discussed in more detail in chapters on opportunities in new energy, new materials, and the life sciences.

5.6 *Crosscutting Issues: Engineering Simulation*

Simulation and modeling are integral to every engineering activity. While the use of simulation and modeling has been widespread in engineering, the WTEC study found several major hurdles still present in the effective use of these tools.

Finding 1: Software and data interoperability, visualization, and algorithms that outlast hardware obstruct more effective use of engineering simulation.

Interoperability of software and data is a major hurdle resulting in limited use of simulation software by non-simulation experts. Commercial vendors are defining de facto standards, and there is little effort beyond syntactic compatibility. Codes are too complicated to permit any user customization, and effective workflow methods need to be developed to aid in developing simulations for complex systems.

- In most engineering applications, algorithms, software and data/visualization are primary bottlenecks. Computational resources (flops and bytes) were not limiting factors at most sites. Lifecycle of algorithms is in the 10–20 years range, whereas hardware lifecycle is in the 2–3 years range. Visualization of simulation outputs remains a challenge and HPC and high-bandwidth networks have exacerbated the problem.

Finding 2: Links between physical and system level simulations remain weak. There is little evidence of atom-to-enterprise models that are coupled tightly with process and device models and thus an absence of multiscale SBE&S to inform strategic decision-making directions.

Experimental validation of models remains difficult and costly, and uncertainty quantification is not being addressed adequately in many of the applications. Models are often constructed with insufficient data or physical measurements, leading to large uncertainty in the input parameters. The economics of parameter estimation and model refinement are rarely considered, and most engineering analyses are conducted under deterministic settings. Current modeling and simulation methods work well for existing products and are mostly used to understand/explain experimental observations. However, they are not ideally suited for developing new products that are not derivatives of current ones.

Finding 3: Although U.S. academia and industry are, on the whole, ahead (marginally) of their European and Asian counterparts in the use of engineering simulation, pockets of excellence exist in Europe and Asia that are more advanced than U.S. groups, and Europe is leading in training the next generation of engineering simulation experts.

- Examples of pockets of excellence in engineering simulation include Toyota, Airbus, and the University of Stuttgart.
- European and Asian researchers rely on the United States to develop the common middleware tools, whereas their focus is on application-specific software.
- The transition from physical systems modeling to social-scale engineered systems in the United States lags behind the Japanese, who are, e.g., modeling the behavioral patterns of six billion people using the Life Simulator.
- European universities are leading the world in developing curricula to train the next generation of engineering simulation experts, although the needed combination of domain, modeling, mathematical, computational and decision-making skills is still rare.

5.7 *Crosscutting Issues: Validation, Verification, and Uncertainty Quantification*

The NSF SBES report (Oden *et al.*, 2006) stresses the need for new developments in V&V and UQ in order to increase the reliability and utility of the simulation methods in the future at a profound level. A report on European computational science (ESF, 2007) concludes that "without validation, computational data are not credible, and hence, are useless." The aforementioned National Research Council report on integrated computational materials engineering (ICME) (NRC, 2008) states that, "Sensitivity studies, understanding of real world uncertainties, and experimental validation are key to gaining acceptance for and value from ICME tools that are less than 100 percent accurate." A clear recommendation was reached by a recent study on Applied Mathematics by the U.S. Department of Energy (Brown, 2008) to "significantly advance the theory and tools for quantifying the effects of uncertainty and numerical simulation error on predictions using complex models and when fitting complex models to observations."

The data and other information the WTEC panel collected in its study suggests that there are a lot of "simulation-meets-experiment" types of projects but no systematic effort to establish the rigor and the requirements on UQ and V&V that the cited reports have suggested are needed.

Finding 1: Overall, the United States leads the research efforts today, at least in terms of volume, in quantifying uncertainty,

mostly in computational mechanics; however, there are similar recent initiatives in Europe.

- Specifically, fundamental work in developing the proper stochastic mathematics for this field, e.g., in addressing the high-dimensionality issue, is currently taking place in Germany, Switzerland, and Austria.

In the United States, the DOD Defense Modeling and Simulation Office (DMSO) has been the leader in developing V&V frameworks, and more recently the Department of Energy has targeted UQ through the ASCI program and its extensions (including its current incarnation, PSAAP). The ASCI/PSAAP program is focused on computational physics and mechanics problems, whereas DMSO has historically focused on high-level systems engineering, e.g., warfare modeling and simulation-based acquisition. One of the most active groups in V&V and UQ is the Uncertainty Estimation Department at Sandia National Labs, which focuses mostly on quantifying uncertainties in complex systems. However, most of the mathematical developments are taking place in universities by a relatively small number of individual researchers.

Finding 2: Although the U.S. Department of Defense and Department of Energy have been leaders in V&V and UQ efforts, they have been limited primarily to high-level systems engineering and computational physics and mechanics, respectively, with most of the mathematical developments occurring in universities by small numbers of researchers. In contrast, several large European initiatives stress UQ-related activities.

There are currently no funded U.S. national initiatives for fostering collaboration between researchers who work on new mathematical algorithms for V&V/UQ frameworks and design guidelines for stochastic systems. In contrast, there are several European initiatives within the Framework Programs to coordinate research on new algorithms for diverse applications of computational science, from nanotechnology to aerospace. In Germany, in particular, UQ-related activities are of utmost importance at the Centers of Excellence; for example, the University of Stuttgart has a chaired professorship of UQ (sponsored by local industry), and similar initiatives are taking place in the Technical University of Munich and elsewhere.

Finding 3: Existing graduate level curricula, worldwide, do not teach stochastic modeling and simulation in any systematic way.

Indeed, very few universities offer regular courses in stochastic partial differential equations (SPDEs), and very few textbooks exist on numerical solution of SPDEs. The typical graduate coursework of an engineer does not include advanced courses in probability or statistics, and important topics such as design or mechanics, for example, are taught with a purely deterministic focus. Statistical mechanics is typically taught in Physics or Chemistry departments that may not fit the background or serve the requirements of students in Engineering. Courses in mechanics and other disciplines that emphasize a statistical description *at all scales* (and not just at the small scales) due to both intrinsic and extrinsic stochasticity would be particularly effective in setting a solid foundation for educating a new cadre of simulation scientists.

5.8 *Crosscutting Issues: Multiscale Modeling and Simulation*

True multiscale simulation has long been a goal of SBE&S. It naturally arises as a desirable capability in virtually every application area of SBE&S. It particularly arises when the notion of design and/or control comes into play, since in this case the impact of changes in problem definition at the larger scales needs to be understood in terms of the impact on structures at the smaller scales. For example, in materials, the question might be how changes in the desirable performance characteristics of a lubricant translate into differences in molecular structure of the constituent molecules, so that new lubricants can be synthesized. In biology, the question might be what needs to happen at the cell–cell interaction level in order to control the spread of cancer. In energy and sustainability, the goal might be to understand how setting emissions targets for carbon dioxide at power plants in the United States may affect the growth and biodiversity of South American rain forests. Answering all of these questions requires a combination of downscaling and upscaling, and so fall in the realm of multiscale simulation.

Finding 1: Multiscale modeling is exceptionally important and holds the key to making SBE&S more broadly applicable in areas such as design, control, optimization, and abnormal situation modeling. However, although successful examples exist within narrow disciplinary boundaries, attempts to develop general strategies have not yet succeeded.

Finding 2: The lack of code interoperability is a major impediment to industry's ability to link single-scale codes into a multiscale framework.

Finding 3: Although U.S. research in multiscale simulation today is on a par with Japan and Europe, it is diffuse, lacking focus and integration, and federal agencies have not traditionally supported the long-term development of codes that can be distributed, supported, and successfully used by others. This contrasts starkly with efforts in Japan and Europe, where large, interdisciplinary teams are supported long term to distribute codes either in open-source or commercial form.

- The current standing of the United States in multiscale modeling and simulation is due in part to U.S. leadership in high-performance computing resources. This is because to validate a multiscale simulation methodology it is necessary to perform the simulation at full detail enough times to provide a validation data set, and this may require enormous computational resources. Thus, continued pre-eminence in leadership-class computing may underpin future U.S. leadership in multiscale modeling and simulation.
  - One concern noted by the panel is that access to the largest computational facilities (such as the emerging petascale platforms) may be difficult for a project whose stated aim is to run a series of petaflop-level simulations in order to develop and validate a multiscale modeling methodology. Access to petaflop-level resources today is generally focused on solving a small set of problems, as opposed to developing the methodologies that will enable the solution of a broader range of problems.

5.9 *Crosscutting Issues: Big Data, Visualization, and Data-driven Simulation*

The role of big data and visualization in driving new capabilities in SBE&S is pronounced and critical to progress in the three thematic areas of this report. We can also interpret the "big data" challenges faced by the SBE&S community as a recursive resurfacing of the CPU-memory "data bottleneck" paradigm from the chip architecture scale to societal scale. The locus of large-scale efforts in data management correlates by discipline and not by geographical region.

Finding 1: The biological sciences and the particle physics communities are pushing the envelope in large-scale data management and visualization methods. In contrast, the chemical

and material science communities lag in prioritization of investments in the data infrastructure.

- In the biological sciences, in both academic laboratories and industrial R&D centers, there is an appreciation of the importance of integrated, community-wide infrastructure for dealing with massive amounts of data, and addressing issues of data provenance, heterogeneous data, analysis of data and network inference from data. There are great opportunities for the chemical and materials communities to move in a similar direction, with the promise of huge impacts on the manufacturing sector.

Finding 2: Industry is significantly ahead of academia with respect to data management infrastructure, supply chain, and workflow.

- Even within a given disciplinary area, there is a notable difference between industrial R&D centers and academic units, with the former placing significantly more attention to data management infrastructure, data supply chain, and workflow, in no small part due to the role of data as the foundation for intellectual property (IP) assets. In contrast, most universities lack a campus-wide strategy for the ongoing transition to data-intensive research and there is a widening gap between the data infrastructure needs of the current generation of graduate students and the capabilities of the campus IT infrastructure. Moreover, industrial firms are particularly active and participate in consortia to promote open standards for data exchange — a recognition that SBE&S is not a series of point solutions but an integrated set of tools that form a workflow engine. Companies in highly regulated industries, e.g., biotechnology and pharmaceutical companies, are also exploring open standards and data exchange to expedite the regulatory review processes for new products.

Finding 3: Big data and visualization capabilities are inextricably linked, and the coming "data tsunami" made possible by petascale computing will require more extreme visualization capabilities than are currently available, as well as appropriately trained students who are proficient in data infrastructure issues.

- Big data and visualization capabilities go hand in hand with community-wide software infrastructure; a prime example is in the particle physics community and the Large Hadron Collider infrastructure networking

CERN to the entire community in a multi-tier fashion. Here, visualization techniques are essential given the massive amounts of data, and the rarity of events (low signal to noise ratio) in uncovering new science.

Finding 4: Big data, visualization and dynamic data-driven simulations are crucial technology elements in numerous "grand challenges," including the production of transportation fuels from the last remaining giant oil fields.

- Global economic projections for the next two decades and recent examples of price elasticity in transportation fuels suggest that the scale of fluctuation in reservoir valuations would be several orders of magnitude beyond the global spending in SBE&S research.

5.10 *Crosscutting Issues: Education and Training*

Education and training of the next generation of computational scientists and engineers proved to be the number one concern at nearly all of the sites visited by the panel. A single, overarching major finding can be summarized as follows:

Finding: Continued progress and U.S. leadership in SBE&S and the disciplines it supports are at great risk due to a profound and growing scarcity of appropriately trained students with the knowledge and skills needed to be the next generation of SBE&S innovators.

This is particularly alarming considering that, according to numerous recent reports,

- The U.S. lead in most scientific and engineering enterprises is decreasing across all S&E indicators.
- The number of U.S. doctorates both in the natural sciences and in engineering has been surpassed by those in the European Union and by those in Asia.
- Countries in the European Union and Japan are scampering to recruit foreign students to make up for the dwindling numbers of their youths. This is driving fierce competition for international recruiting. The United States is sitting at exactly the "fertility replacement rate" and only a small percentage of its population joins scientific and engineering

professions. China and India, on the other hand, are likely to become increasingly strong SBE&S competitors.

The panel observed many indicators of increased focus on SBE&S education in China and in Europe:

- The European Union is investing in new centers and programs for education and training in SBE&S — all of an interdisciplinary nature. New BSc and MSc degrees are being offered in SBE&S through programs that comprise a large number of departments. In many cases, a complete restructuring of the university has taken place in order to create, for instance, MSc degrees, graduate schools, or international degrees in simulation technology.
- European and Asian educational/research centers are attracting more international students (even from the United States). Special SBE&S programs (in English) are being created for international students. The United States is seeing smaller increases in international enrollment compared to other countries. This is due not only to post-9/11 security measures but also — and increasingly — to active competition from other countries.

All countries, however, share many of the same challenges when it comes to SBE&S education and training:

- Students are trained primarily to run existing codes rather than to develop the skills in computational mathematics and software engineering necessary for the development of the next generation of algorithms and software. There are pitfalls in interdisciplinary education such as computational science and engineering, including a tradeoff between breadth and depth. In order to solve "grand challenge" problems in a given field, solid knowledge of a core discipline, in addition to computational skills, is crucial.
- Demand exceeds supply. There is a huge demand in the European Union and Asia for qualified SBE&S students who get hired immediately after their MSc degrees by industry or finance: there is both collaboration and competition between industry and academia. This phenomenon in the United States may be tempered in the future due to the current economic situation. However, one might argue that the need for increased physics-based modeling and simulation in macroeconomics and financial economics, and in industry in general as it cuts costs while preserving the ability for innovation, has never been more urgent.

The WTEC Panel on Simulation-Based Engineering and Science
S.C. Glotzer (Chair), S.T. Kim (Vice Chair), P.T. Cummings,
A. Deshmukh, M. Head-Gordon, G. Karniadakis, L. Petzold,
C. Sagui and M. Shinozuka
April 2009

References

Benioff, M. R. and Lazowska, E. D. (2005). *Computational Science: Ensuring America's Competitiveness* (President's Information Technology Advisory Committee, Washington, DC), http://www.nitrd.gov/pitac/reports/20050609_computational/computational.pdf.

European Computational Science Forum of the European Science Foundation (2007). *The Forward Look Initiative. European Computational Science: The Lincei Initiative: From Computers to Scientific Excellence* (European Science Foundation, Strasbourg, France), http://www.esf.org/activities/forward-looks/all-current-and-completed-forward-looks.html.

European Strategy Forum for Research Infrastructures (2006). *European Roadmap for Research Infrastructures: Report 2006* (Office for Official Publications of the European Communities, Luxembourg), ftp://ftp.cordis.europa.eu/pub/esfri/docs/esfri-roadmap-report-26092006_en.pdf.

Oden, J. T., Belytschko, T., Hughes, T. J. R., Johnson, C., Keyes, D., Laub, A., Petzold, L., Srolovitz, D. and Yip, S. (2006). *Revolutionizing Engineering Science through Simulation: A Report of the National Science Foundation Blue Ribbon Panel on Simulation-Based Engineering Science* (National Science Foundation, Arlington, VA), http://www.nsf.gov/pubs/reports/sbes_final_report.pdf.

White House Office of Management and Budget (2008). *Agency NITRD Budget by Program Component Area FY 2008 Budget Estimate and Budget Requests. President's FY 2009 Budget* (OMB, Washington, DC).

National Research Council (2008). *National Academy of Engineering Report of Committee on Integrated Computational Materials Engineering (ICME)* (National Academies Press, Washington, DC).

Further Reading

Atkins, D. M. (2006). *NSF's Cyberinfrastructure Vision for 21st Century Discovery*, Version 7.1 (National Science Foundation, Washington, DC).

Bramley, R., Char, B., Gannon, D., Hewett, T., Johnson, C. and Rice, J. (2000). Enabling technologies for computational science: frameworks, middleware and environments, *Proc. Wkp on Scientific Knowledge, Information, and Computing*, eds. Houstis, E., Rice, J., Gallopoulos, E. and Bramley, R. (Kluwer Academic Publishers, Boston), pp. 19–32.

Committee on Science, Engineering, and Public Policy (2007). *Rising Above the Gathering Storm: Energizing and Employing America for a Brighter Economic Future. Committee on Prospering in the Global Economy of the 21st Century: An Agenda for American Science and Technology* (National Academies Press, Washington, DC), http://www.nap.edu/catalog.php?record_id=11463#orgs.

Council on Competitiveness (2004). *Final Report from the High Performance Users Conference: Supercharging U.S. Innovation and Competitiveness* (Council on Competitiveness, Washington, DC).

Coveney, P., *et al.* (2006). *UK Strategic Framework for High-End Computing*, http://www.epsrc.ac.uk/CMSWeb/Downloads/Other/2006HECStrategicFramework.pdf.

Department of Defense (2002). *Report on High Performance Computing for the National Security Community*, http://www.hpcmo.hpc.mil/Htdocs/documents/04172003_hpc_report_unclass.pdf.

Department of Defense, Defense Science Board (2000), *Report of the Defense Science Board Task Force on DoD Supercomputing Needs*, http://www.acq.osd.mil/dsb/reports/dodsupercomp.pdf.

Department of Energy, Office of Science (2000). *Scientific Discovery through Advanced Computing.*

Executive Office of the President, Office of Science and Technology Policy (2004). *Federal Plan for High-End Computing: Report of the High-End Computing Revitalization Task Force.*

Interagency Panel on Large Scale Computing in Science and Engineering (1982). *Report of the Panel on Large Scale Computing in Science and Engineering* (Department of Defense, National Science Foundation, Department of Energy, and National Aeronautics and Space Administration, Washington, DC).

Joy, W. and Kennedy, K., eds. (1999). *President's Information Technology Advisory Committee Report to the President: Information Technology Research: Investing in Our Future*, http://www.nitrd.gov/pitac/report/.

Joseph, E., Snell, A. and Willard, C. G. (2004). *Council on Competitiveness Study of U.S. HPC Users*, http://www.compete.org/pdf/HPC_Users_Survey.pdf.

Kaufmann, N. J., Willard, C. G., Joseph, E. and Goldfarb, D. S. (2003). *Worldwide High Performance Systems Technical Computing Census.* IDC Report No. 62303.

Keyes, D., Colella, P., Dunning, T. H., Jr. and Gropp, W. D., eds. (2003). *A Science-Based Case for Large-Scale Simulation*, vol. 1 (DOE Office of Science, Washington, DC), http://www.pnl.gov/scales/.

Keyes, D., Colella, P., Dunning, T. H., Jr. and Gropp, W. D., eds. (2004). *A Science-Based Case for Large-Scale Simulation*, vol. 2 (DOE Office of Science, Washington, DC), http://www.pnl.gov/scales/.

National Research Council, Computer Science and Telecommunications Board (2005). *Getting Up to Speed: the Future of Supercomputing* (National Academies Press, Washington, DC).

National Research Council (2008). *The Potential Impact of High-End Capability Computing on Four Illustrative Fields of Science and Engineering* (National Academies Press, Washington, DC).

National Science Foundation Blue Ribbon Panel on High Performance Computing (1993). *From Desktop to Teraflop: Exploiting the U.S. Lead in High Performance Computing.*

National Science Foundation (1995). *Report of the Task Force on the Future of the NSF Supercomputer Centers Program.*

National Science Foundation (2003). *Revolutionizing Science and Engineering through Cyberinfrastructure: Report of the National Science Foundation Blue Ribbon Advisory Panel on Cyberinfrastructure.*

President's Council of Advisors on Science and Technology (2004). *Report to the President: Sustaining the Nation's Innovation Ecosystems, Information Technology, Manufacturing, and Competitiveness.*

Society for Industrial and Applied Mathematics Working Group on Computational Science and Engineering (CSE) Education (2001). Graduate education in computational science and engineering, *SIAM Review* 43(1), 163–177.

Workshop and Conference on Grand Challenges Applications and Software Technology (1993). Pittsburgh, Pennsylvania.

Chapter 1

INTRODUCTION

Sharon C. Glotzer

1.1 Background and Scope

The impetus for this study builds upon many national and international reports highlighting the importance and promise of computational science and engineering. From the 1982 US interagency study on computational science now known as the Lax Report (Lax, 1982) to the 2006 U.S. National Science Foundation Blue Ribbon panel study on simulation-based engineering science known as the Oden report (Oden *et al.*, 2006) to the 2008 National Research Council Reports on High End Capability Computing for Science & Engineering (NRC, 2008a) and Integrated Computational Materials Engineering (NRC, 2008b), critical advances in such fields as medicine, nanotechnology, aerospace and climate prediction made possible by application of high-end computing to the most challenging problems in science and engineering have been well documented. Yet, as demonstrated in the 2005 report of the President's Information Technology Advisory Committee (PITAC) (Benioff and Lazowska, 2005), investment in computational science in the United States and the preparation of the next generation of computational researchers remains insufficient to fully leverage the power of computation for solving the biggest problems that face us as a nation going forward. As the United States and Japan continue to invest in the next generation of computer architectures, with the promise of thousand-fold or more increases of computer power coming in the next half-decade, an assessment of where the United States stands relative to other countries in its use of, commitment to, and leadership in computer simulation for engineering and science is therefore of urgent interest. That is the focus of this report.

Today we are at a "tipping point" in computer simulation for engineering and science. Computer simulation is more pervasive today — and having more impact — than at any other time in human history. No field of science or engineering exists today that has not been advanced by, and in some cases utterly transformed by, computer simulation. The Oden report (Oden *et al.*, 2006) summarizes the state of simulation-based engineering science today: "*... advances in mathematical modeling, in computational algorithms, and the speed of computers and in the science and technology of data-intensive computing have brought the field of computer simulation to the threshold of a new era, one in which unprecedented improvements in the health, security, productivity, and competitiveness of our nation may be possible.*" The report further argues that a host of critical technologies are on the horizon that cannot be understood, developed, or utilized without simulation methods. Simulation has today reached a level of predictiveness that it now firmly complements the traditional pillars of science and engineering, namely theory and experimentation/observation. Moreover, there is abundant evidence that computer simulation is critical to scientific discovery and engineering innovation. At the same time, computers are now affordable and accessible to researchers in every country around the world. The near-zero entry-level cost to perform a computer simulation means that anyone can practice SBE&S, and from anywhere. Indeed, the world of computer simulation is flattening, and becoming flatter every day. In that context, it is therefore meaningful to examine the question of what it means for any nation, the United States in particular, to lead in SBE&S.

The next decade is shaping up to be a transformative one for SBE&S. A new generation of massively multicore computer chip architectures is on the horizon that will lead to petaflop computing, then exaflop computing, and beyond. These architectures will allow unprecedented accuracy and resolution, as well as the ability to solve the highly complex problems that face society today. The toughest technological and scientific problems facing society, from finding alternative energy sources to global warming to sustainable infrastructures, to curing disease and personalizing medicine, are *big* problems. They are complex and messy, and their solution requires a partnership among experiment, theory, and simulation working across all of the disciplines of science and engineering. Recent reports like *Rising Above the Gathering Storm* (COSEPUP, 2007) argue that our nation is at risk of losing its competitive edge. Our continued capability as a nation to lead in simulation-based discovery and innovation is key to our ability to compete in the twenty-first century.

Yet because it is often viewed more as an enabling technology for other disciplines, rather than a discipline in its own right, investment in and support of computational science and engineering is often not prioritized as it should be at all levels of the R&D enterprise. The 2005 PITAC (Benioff and Lazowska, 2005) report writes, "*The universality of computational science is its intellectual strength. It is also its political weakness. Because all research domains benefit from computational science but none is solely defined by it, the discipline has historically lacked the cohesive, well-organized community of advocates found in other disciplines. As a result, the United States risks losing its leadership and opportunities to more nimble international competitors. We are now at a pivotal point, with generation-long consequences for scientific leadership, economic competitiveness, and national security if we fail to act with vision and commitment.*"

To provide program managers in U.S. research agencies as well as researchers in the field a better understanding of the status and trends in SBE&S R&D abroad, in 2007 the National Science Foundation (NSF), in cooperation with the Department of Energy (DOE), NASA, National Institutes of Health (NIH), National Library of Medicine, the National Institute of Standards and Technology (NIST), and the Department of Defense (DOD), sponsored this *WTEC International Assessment of Simulation-Based Engineering and Science.* The study was designed to gather information on the worldwide status and trends in SBE&S research and to disseminate it to government decision makers and the research community. The study participants reviewed and assessed the state of the art in SBE&S research and its application in academic and industrial research. Questions of interest to the sponsoring agencies addressed by the study included:

- Where are the next big breakthroughs and opportunities coming from in the United States and abroad, and in what fields are they likely to be?
- Where is the United States leading, trailing, or in danger of losing leadership in SBE&S?
- What critical investments in SBE&S are needed to maintain U.S. leadership, and how will those investments impact our ability to innovate as a nation?

Simulation-based engineering and science is a vast field, spanning the physical, chemical, biological, and social sciences and all disciplines of engineering, and it would be impossible to cover all of it in any meaningful way. Instead, the study focused on SBE&S and its role in three primary thematic areas that will be transformative in the health and prosperity, and in the economic and national security, of the United States. The thematic areas studied are:

- Life sciences and medicine;
- Materials; and
- Energy and sustainability.

In turn, these domains are ones in which SBE&S promises to have a transformative effect in the near term provided that continued and, in many cases, expedited progress occur in the following crosscutting areas, which were identified in the Oden report (Oden *et al.*, 2006) and included in this study:

- Next generation architectures and algorithms
- Scientific and engineering simulation software development
- Engineering simulation
- Validation, verification and uncertainty quantification
- Multiscale modeling and simulation
- Big data, visualization, and data-driven simulation
- Education and training

As with all WTEC reports, this report will present findings that can be used as guidance for future recommendations and investments.

1.2 Methodology

The agency sponsors recruited the study chair, Sharon C. Glotzer, Professor of Chemical Engineering, Materials Science and Engineering, Physics, Applied Physics, and Macromolecular Science and Engineering at the University of Michigan, Ann Arbor, in early 2007. On March 13, 2007 an initial meeting was held at NSF headquarters in Arlington, VA with the agency sponsors, WTEC representatives, and the study chair, to establish the scope of the assessment. WTEC then recruited a panel of U.S. experts (see Table 1.1).[1] The assessment was initiated by a kickoff meeting on

[1]The panel is grateful to panelist Masanobu Shinozuka who, despite being unable to travel to the U.S. Baseline Workshop and the international trips due to an injury, participated in the initial and final workshops at NSF and contributed the chapter on Energy & Sustainability in the final report.

Table 1.1. Panel members.

Panelist	Affiliation
Sharon C. Glotzer (Panel Chair)	University of Michigan, Ann Arbor
Sangtae Kim (Vice Chair)	Morgridge Institute for Research
Peter Cummings	Vanderbilt University and Oak Ridge National Laboratory
Abhijit Deshmukh	Texas A&M University
Martin Head-Gordon	University of California, Berkeley
George Karniadakis	Brown University
Linda Petzold	University of California, Santa Barbara
Celeste Sagui	North Carolina State University
Masanobu Shinozuka	University of California, Irvine

Table 1.2. Study advisors.

Panelist	Affiliation
Jack Dongarra	University of Tennessee and Oak Ridge, National Laboratory
James Duderstadt	University of Michigan, Ann Arbor
J. Tinsley Oden	University of Texas, Austin
Gilbert S. Omenn	University of Michigan, Ann Arbor
Tomás Diaz de la Rubia	Lawrence Livermore National Laboratory
David Shaw	D. E. Shaw Research and Columbia University
Martin Wortman	Texas A&M University

July 10, 2007 at NSF headquarters in Arlington, VA. Participants discussed the scope of the project and the need for a North American baseline workshop, candidate sites in Europe and Asia for panel visits, the overall project schedule, and assignments for the final report. WTEC also recruited, at the request of the chair, a team of advisors (Table 1.2) to provide additional input to the process. Brief biographies of the panelists and advisors are presented in Appendix A.

The panelists, sponsors, and WTEC convened a U.S. Baseline Workshop on November 20–21, 2007 at the Hilton Hotel in Arlington, VA to report on the current status of U.S. R&D in simulation-based engineering & science. Table 1.3 lists the speakers and the titles of their presentations.

The speakers at the U.S. Baseline Workshop were selected by the panelists with input from the advisors and program managers, and with an eye towards complementing the expertise and knowledge of the study panel and emphasizing the thematic areas and crosscutting issues chosen

Table 1.3. Speakers at the U.S. baseline workshop.

Name	Affiliation	Presentation title
Michael Reischman	NSF	Importance of Field to NSF
Duane Shelton	WTEC	Context of Study
Sharon Glotzer (Chair)	University of Michigan	Introduction and Study Overview
J. Tinsley Oden	University of Texas at Austin	The NSF Blue Ribbon Panel Report on SBES
Peter Cummings	Vanderbilt University	Materials (Session Introduction)
Emily Carter	Princeton University	Status & Challenges in Quantum Mechanics Based Simulations of Materials Behavior
Tomás Diaz de la Rubia (presented by Tom Arsenlis)	Lawrence Livermore National Laboratory	Toward Petaflops Computing for Materials, National Security, Energy, and Climate Applications
Sangtae Kim (Vice Chair) for M. Shinozuka	Morgridge Institute for Research	Energy & Sustainability (Session Introduction)
Roger Ghanem	University of Southern California	Sustainability by Design: Prediction and Mitigation of Complex Interactions in the Urban Landscape
Lou Durlofsky	Stanford University	Computational Challenges for Optimizing Oil and Gas Production and CO_2 Sequestration
R.T. (Rick) Mifflin (presented by Sangtae Kim)	Exxon Mobil	High-Performance Computing and U.S. Competitiveness in the Practice of Reservoir Simulation
Linda Petzold	University of California, Santa Barbara	Life Sciences (Session Introduction)
David Shaw	D. E. Shaw Research	Molecular Dynamics Simulations in the Life Sciences
Alex Bangs	Entelos	Predictive Biosimulation in Pharmaceutical R&D
Peter Cummings	Vanderbilt University	Multiscale Simulation (Session Introduction)
Ioannis Kevrekidis	Princeton University	Equation-Free Modeling and Computation for Complex/Multiscale Systems
W.K. Liu	Northwestern University	SBE&S Approach to Analysis and Design of Microsystems: From a Dream to a Vision to Reality
Abhi Deshmukh	Texas A&M University	Engineering Design (Session Introduction)

(Continued)

Table 1.3. (*Continued*)

Name	Affiliation	Presentation title
Troy Marusich	Third Wave Technologies	Improved Machining via Physics-Based Modeling
Karthik Ramani	Purdue University	Leveraging Simulations for Design: Past-Present-Future
George Karniadakis	Brown University	Validation, Verification, and Uncertainty Quantification (Session Introduction and Review)
Sangtae Kim (Vice Chair)	Morgridge Institute for Research	Big Data (Session Introduction)
Fran Berman	San Diego Supercomputer Center	Simulation and Data
Ernst Dow	Eli Lilly	Big Data and Visualization
Linda Petzold	University of California, Santa Barbara	Next-Generation Algorithms/High-Performance Computing (Session Introduction)
Jack Dongarra	University of Tennessee	An Overview of High Performance Computing and Challenges for the Future
William D. Gropp	Argonne National Laboratory	Architecture Trends and Implications for Algorithms
David Keyes	Columbia University	Attacking the Asymptotic Algorithmic Bottleneck: Scalable Solvers for SBE&S
Martin Head-Gordon	University of California, Berkeley	Software Development (Session Introduction)
Dennis Gannon	Indiana University	Software for Mesoscale Storm Prediction: Using Supercomputers for On-the-Fly Tornado Forecasting
Chau-Chyun Chen	Aspen Institute	High Impact Opportunities in Simulation-Based Product and Process Design and Development
Celeste Sagui	North Carolina State University	Education (Session Introduction)
J. Tinsley Oden	University of Texas at Austin	Goals and Barriers to Education in SBE&S

for the study. The workshop included ample time for discussion. To kick off the workshop, Michael Reischman of NSF and Duane Shelton of WTEC provided welcoming remarks, discussed the importance of simulation-based engineering and science to NSF, and provided the context for the study.

Study chair Sharon Glotzer presented an overview of the planned study and study process. Panel advisor J. Tinsley Oden of the University of Texas, Austin, who chaired the NSF Blue Ribbon Panel Study on Simulation-Based Engineering Science, provided a summary of that study.

Following this introduction, sessions were organized to target the specific themes and crosscutting issues chosen for the study. In the Materials session, chaired by panelist Peter Cummings, Emily Carter of Princeton University discussed the status and challenges of quantum-mechanics (QM) based materials simulations. She provided a summary of open source and commercial QM-based materials simulation codes being developed and/or used today, demonstrating the dearth of QM code development in the United States today relative to other countries and discussing the implications of this. Tom Arsenlis of Lawrence Livermore National Laboratory presented a talk for Tomás Diaz de la Rubia on the path towards petaflop/s computing for materials, national security, energy and climate applications. Arsenlis demonstrated several achievements by the DOE labs in pushing the boundaries of large scale computing. In the session on Energy and Sustainability, chaired by study vice-chair Sangtae Kim, Roger Ghanem of the University of Southern California described how computer simulation is used to predict complex interactions in urban development and design strategies for mitigating unwanted interactions. Lou Durlofsky of Stanford University described the use of computer simulation for optimizing oil and gas production and CO_2 sequestration, and discussed challenges and opportunities for the United States. Sangtae Kim presented a talk prepared by Rick Mifflin of Exxon-Mobil on high performance computing and U.S. competitiveness in the practice of oil reservoir simulation, demonstrating the enormous possibilities that remain for predicting high yield sites with increased use of simulation-based engineering and science. In the Life Sciences session, chaired by panelist Linda Petzold, David Shaw of D.E. Shaw Research demonstrated enormous speedups now possible in molecular dynamics simulation of biomolecules from critical advances in hardware, algorithms, and software. These recent advances now make possible the simulation of biological processes on the relevant time scales over which they occur. Alex Bangs of Entelos described the use of predictive biosimulation in pharmaceutical R&D, a major growth area for simulation.

In the session on Multiscale Simulation, chaired by panelist Peter Cummings, Ioannis Kevrekidis of Princeton University and W.K. Liu of Northwestern University described the status of method development for bridging time and length scales in complex science and engineering problems, comparing and contrasting efforts in the United States and abroad. In the session on Engineering Design, chaired by panelist Abhi Deshmukh, Troy Marusich of Third Wave Technologies discussed how advances in computational capabilities are now permitting the use of physics-based models in simulations to improve machining. Karthik Ramani of Purdue University described the use of simulation-based engineering and science for engineering design. Panelist George Karniadakis provided a review of, and led a discussion on, validation, verification and uncertainty quantification. He identified issues and challenges and reviewed U.S. efforts in this area. In the session on Big Data, chaired by study vice-chair Sangtae Kim, Fran Berman of the San Diego Supercomputer Center discussed the enormous data challenges associated with petascale (and beyond) simulations. Ernst Dow of Eli Lilly discussed trends in visualization both in the United States and abroad and the importance of visualization in interpreting and exploring very large data sets.

In the session on Next Generation Algorithms and High Performance Computing, chaired by panelist Linda Petzold, Jack Dongarra of the University of Tennessee presented an overview of high performance computing and its evolution over the years. He compared the power of laptop computers today with room-sized supercomputers of a decade earlier, and discussed future trends in HPC, including the great opportunities and challenges accompanying manycore chip architectures. Bill Gropp of Argonne National Laboratory and David Keyes of Columbia University also discussed architecture trends and described their implications for algorithms. Keyes described the critical importance of algorithms in achieving the many orders of magnitude in speed up needed to tackle some of the most challenging problems in science and engineering.

The session on Software Development was chaired by panelist Martin Head-Gordon of the University of California, Berkeley. In this session, Dennis Gannon of Indiana University described the use of high performance computing for storm prediction and tornado forecasting, and C.C. Chen of the Aspen Institute discussed high impact opportunities in simulation-based product and process design and development. The final session of the Baseline Workshop focused on Education and was chaired by panelist Celeste Sagui. J. Tinsley Oden of the University of Texas, Austin, discussed the need for new organizational structures and approaches within universities to support computational science and engineering education. The session sparked perhaps the most passionate and lengthy discussions of the workshop, on the topic of educating the next generation of computational scientists and engineers. This is a theme that arose throughout the study as one of great concern around the world, as discussed later in this report.

A bibliometric analysis (see Appendix C), combined with input from the panelists, advisors, and participants of the Baseline Workshop and other experts, provided guidance in selecting European and Asian sites to visit. Sites were selected as representative of what were perceived or benchmarked to be the most productive and/or most reputable and/or highest impact research labs and groups outside of the United States in the various areas of SBE&S comprising the study. Because the field is so broad and both time and resources finite, it was impossible to visit every leading group or research lab, or to visit every country where major rapid growth in research, including computational science, is underway (most notably Singapore and Saudi Arabia). We note that more sites were visited for this study than in any previous WTEC study. The panel therefore relied on its own knowledge of international efforts, as well as informal conversations with colleagues and briefings heard in different venues, to complement what we learned in the site visits.

The international assessment phase of the WTEC study commenced in December 2007 with visits to the 21 sites in China and Japan shown in Figs. 1.1, 1.2, and Table 1.4. To focus the discussions in the various laboratories, the host engineers and scientists were provided with a set of questions (Appendix B) compiled by the panel in advance of the visits. While the discussion did not necessarily follow the specific questions, they provided a general framework for the discussions. During its visit to China, the WTEC panel was privileged to attend a symposium on SBE&S organized by Dalian University, at which approximately 75 faculty and students heard presentations from a dozen faculty members in China whose research groups are actively exploiting SBE&S. Our hosts both in Asia and Europe demonstrated a wide range of computational research applied to many important problems in science and engineering. Talks discussed research in bioinformatics, computational nanomechanics, computational fluid dynamics, offshore oil drilling, composite materials, automotive design, polymer science and software engineering, as examples. The Asia trip concluded with a meeting in Nagoya, Japan on December 8, 2007, in which the panelists reviewed and compared their site visit notes. A week of visits

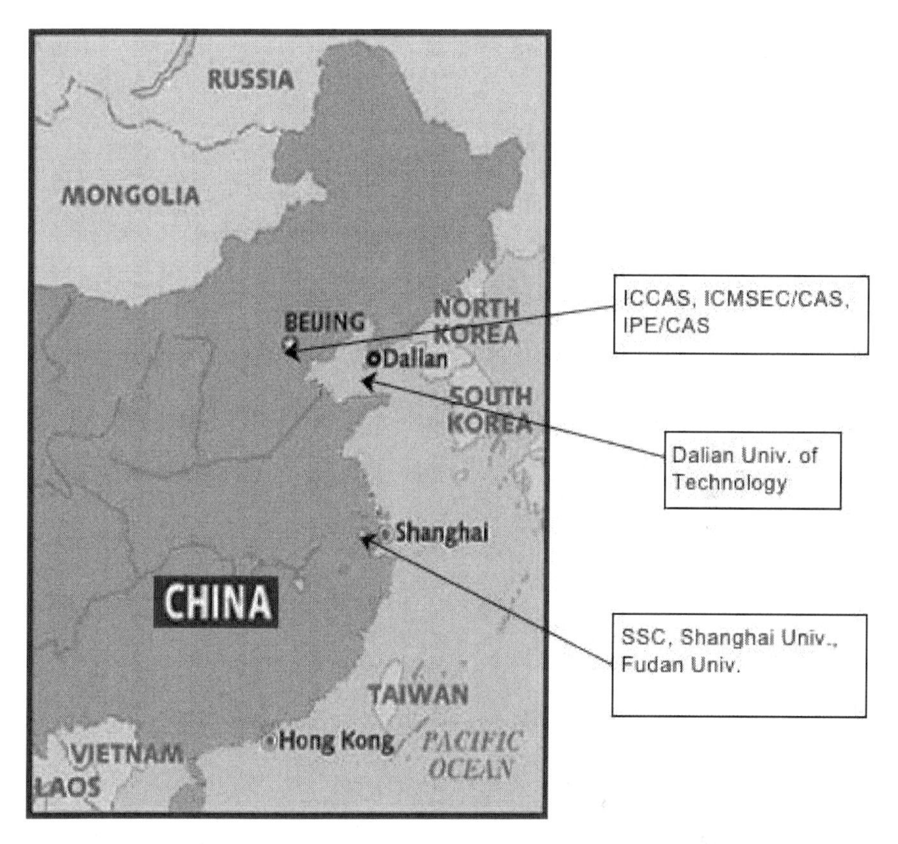

Fig. 1.1. Sites visited in China.

to the 38 European sites listed in Fig. 1.3 and Table 1.5 commenced on February 24, 2008, and concluded with a wrap-up meeting in Frankfurt, Germany on March 1, 2008.

The panel reconvened for a final workshop at NSF on April 21, 2008, to present its findings and conclusions. Presentations focused on the three thematic domains of life sciences and medicine, materials, energy and sustainability, and on the seven crosscutting issues identified above.

1.3 Overview of the Report

This final report closely follows the outline of the final workshop held in April 2008. Following this introductory chapter, we address three major

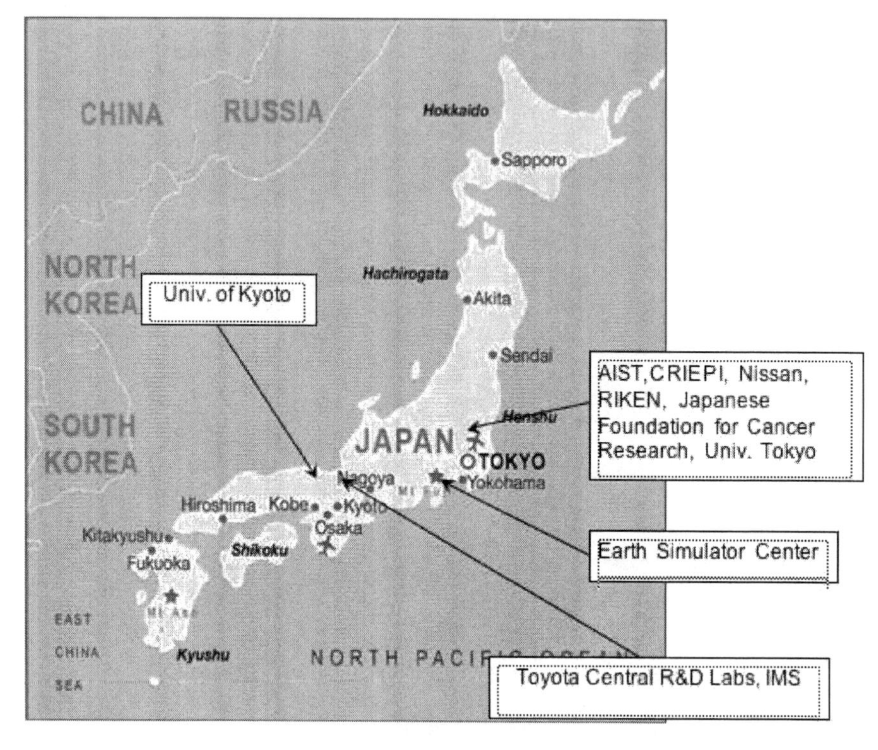

Fig. 1.2. Sites visited in Japan.

thematic domains in which SBE&S is having unprecedented impact and investigate SBE&S research internationally in those domains as compared with the United States. First, Linda Petzold discusses in Chapter 2 how advances in SBE&S are contributing to breakthroughs in the biomedical and life sciences arena. The simulation of biological processes on relevant time and length scales, and of whole organs and complete physiological systems, is becoming possible as we move towards petascale computing. Predictive biosimulation now supports the development of new therapeutic drugs, meeting the pharmaceutical industry's need for new approaches to deal with the complexity of disease and the increased risk and expense of research and development. Biomechanical simulation is used in surgical planning and design of physical therapies, and simulation of the brain is a realistic near-term goal.

In Chapter 3, Peter Cummings describes advances and applications of SBE&S in materials R&D outside the United States. Computational

Table 1.4. Sites visited in Asia.

#	Country	Site	#	Country	Site
1	China	Institute of Process Engineering, Institute of Computational Mathematics and Scientific/Engineering Computing (ICMSEC-CAS)	12	Japan	Central Research Institute-Electric Power Industry (CRIEPI)
2	China	ICCAS, Institute of Chemistry (Joint Lab for Polymer and Materials Science)	13	Japan	Japan Agency for Marine Earth Science and Technology, Earth Simulator Center (ESC)
3	China	Dalian University of Technology, Department of Engineering Mechanics	14	Japan	Japanese Foundation for Cancer Research
4	China	Dalian University of Technology, School of Automobile Engineering	15	Japan	Kyoto University, Department of Synthetic Chemistry and Biological Chemistry
5	China	Fudan University, Yulang Yang Research Group	16	Japan	Nissan
6	China	Peking University, Computer Science and Engineering	17	Japan	Institute for Molecular Science (IMS)
7	China	Shanghai University, School of Material Science and Engineering	18	Japan	Mitsubishi Chemical Group
8	China	Shanghai University, High Performance Computer Center	19	Japan	RIKEN Next-Generation Supercomputer R&D Center
9	China	Tsinghua University, Engineering Mechanics	20	Japan	Toyota Central Research Laboratory
10	Japan	Advanced Industrial Science and Technology (AIST), Research Institute for Computational Science (RICS)	21	Japan	University of Tokyo
11	Japan	AIST, National Institute of Materials Science (NIMS)			

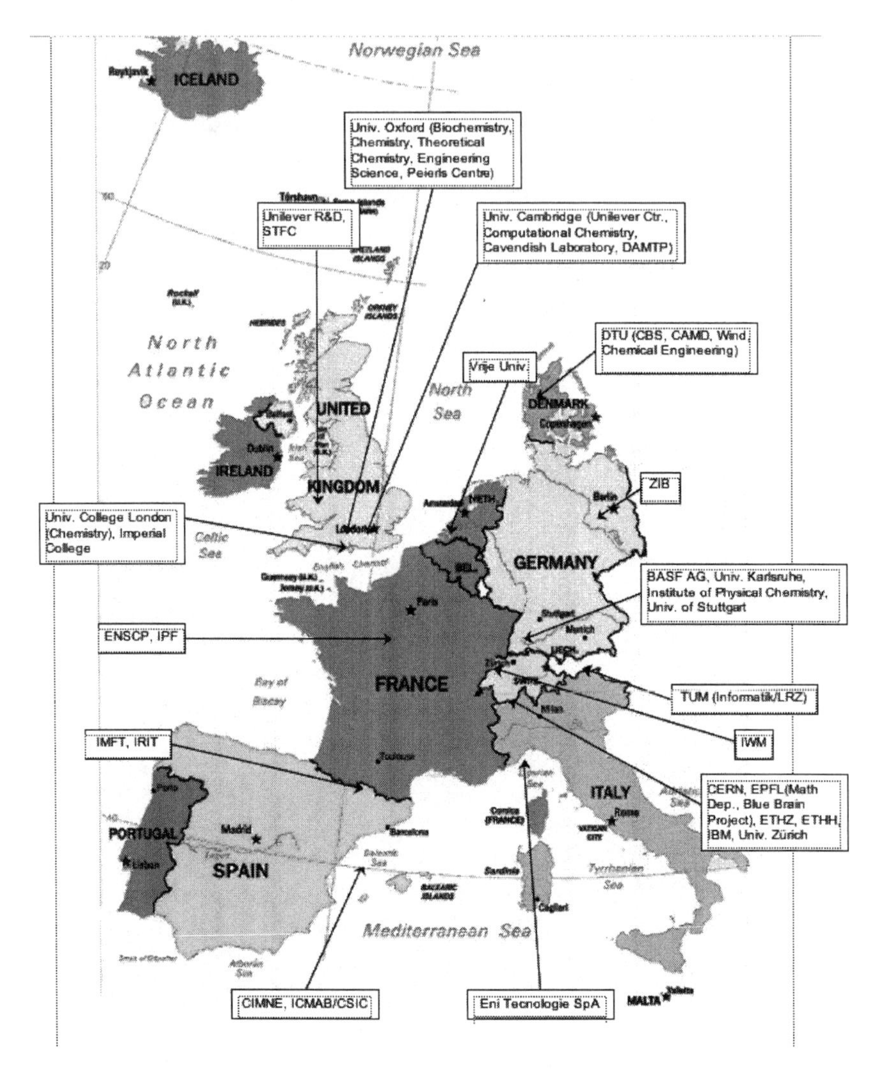

Fig. 1.3. Sites visited in Europe.

materials science and engineering is changing how new materials are discovered, developed, and applied. The convergence of simulation and experiment at the nanoscale is creating new opportunities, acceleration of materials selection into the design process through SBE&S is occurring in some industries, and full-physics simulations are a realistic goal with the increasing power of next generation computers.

Table 1.5. Sites visited in Europe.

#	Country	Site	#	Country	Site
1	Denmark	Technical University of Denmark (DTU), Center for Biological Sequence Analysis, Systems Biology Department (CBS)	20	Switzerland	European Organization for Nuclear Research (CERN)
2	Denmark	DTU, Fluid Mechanics (Wind)	21	Switzerland	Ecole Polytechnique Fédérale de Lausanne (EPFL), Blue Brain Project
3	Denmark	DTU, Physics, Center for Applied Materials Design (CAMD)	22	Switzerland	EPFL (Math and Milan Polytechnic-Quarteroni)
4	Denmark	DTU, Chemical Engineering	23	Switzerland	Swiss Federal Institute of Technology in Zurich (ETHZ) and Hönggerberg (ETHH)
5	France	École Nationale Supérieure de Chimie de Paris (ENSCP)	24	Switzerland	IBM Zürich Research Laboratory (ZRL)
6	France	Institut Français du Pétrole (IFP), Rueil-Malmaison	25	Switzerland	Zürich Universität (Department of Physical Chemistry)
7	France	Institute of Fluid Mechanics (IMFT)	26	United Kingdom	University of Cambridge, Department of Applied Mathematics and Theoretical Physics (DAMTP)
8	France	Institut de Recherche en Informatique (IRIT)	27	United Kingdom	University of Cambridge, Theory of Condensed Matter Group (Cavendish Laboratory)
9	Germany	BASF AG, Ludwigshafen am Rhein	28	United Kingdom	University of Cambridge, Centre for Computational Chemistry
10	Germany	Fraunhofer Institute for the Mechanics of Materials (IWM), Freiburg	29	United Kingdom	University of Cambridge, Unilever Center

(Continued)

Table 1.5. (*Continued*)

#	Country	Site	#	Country	Site
11	Germany	Konrad-Zuse-Zentrum für Informationstechnik Berlin (ZIB)	30	United Kingdom	Science Technology Facilities Council (STFC) (Daresbury Laboratory)
12	Germany	Technical University of Munich and LRZ	31	United Kingdom	Imperial College London, (Rolls Royce UTC/ Thomas Young Centre)
13	Germany	University of Karlsruhe Institute of Physical Chemistry	32	United Kingdom	Unilever Research Laboratory, Bebington
14	Germany	University of Stuttgart, Institute of Thermodynamics and Thermal Process Engineering	33	United Kingdom	University College (Chemistry)
15	Italy	Eni Tecnologie SpA	34	United Kingdom	University of Oxford (Biochemistry, Chemistry, Engineering Science — Fluid Mechanics, Physics)
16	Netherlands	Vrije Universiteit, Molecular Cell Physiology	35	United Kingdom	University of Oxford, Department of Biochemistry
17	Netherlands	Vrije Universiteit, Theoretical Chemistry	36	United Kingdom	University of Oxford, Department of Engineering Science
18	Spain	Universitá Autónoma de Barcelona, Institut de Ciencia de Materials de Barcelona (ICMAB-CSIC)	37	United Kingdom	University of Oxford, Rudolf Peierls Centre for Theoretical Physics
19	Spain	CIMNE (International Center for Numerical Methods in Engineering)	38	United Kingdom	University of Oxford, Physical and Theoretical Chemistry Laboratory

In Chapter 4, Masanobu Shinozuka describe how SBE&S is being used to solve outstanding problems in sustainability and energy. Around the world, SBE&S is aiding in the recovery of untapped oil, the discovery and utilization of new energy sources such as solar, wind, nuclear and biofuels, and the predictive design of sustainable infrastructures. Simulation of multi-billion-agent systems for research in sustainable infrastructures is a realistic goal for petascale computing.

In the succeeding chapters, we present the panel's findings concerning key crosscutting issues that underlie progress and application of SBE&S to important problems in all domains of science and engineering, including the three thematic areas above. In Chapter 5, George Karniadakis highlights trends in new high performance architectures and algorithms driving advances in SBE&S around the world. Massively multicore chip architectures will provide the foundation for the next generation of computers from desktops to supercomputers, promising sustained performance of a petaflop for many scientific applications in just a few years, and exaflops beyond that. Petaflop and exaflop computers under development in the United States and in Japan will be based mainly on concurrency involving, respectively, hundreds of thousands and millions of processing elements with the number of cores per processor doubling every 18–24 months. This disruptive change in computing paradigms will require new compilers, languages, and parallelization schemes as well as smart, efficient algorithms for simplifying computations.

Martin Head-Gordon discusses issues in scientific software development in the United States and abroad in Chapter 6. Advances in software capabilities have revolutionized many aspects of engineering design and scientific research — ranging from computer-aided design in the automotive industry to the development of the chlorofluorocarbon replacements that have helped to begin reversing stratospheric ozone depletion. The significance of these end-use examples is why government support for software development is a good strategic investment — particularly when pursued consistently over the long term.

In Chapter 7, Abhijit Deshmukh describes advances in engineering simulations. Simulation and modeling are integral to every aspect of engineering. Examples of the use of simulation in engineering range from manufacturing process modeling, continuum models of bulk transformation processes, structural analysis, finite element models of deformation and failure modes, computational fluid dynamics for turbulence modeling, multi-physics models for engineered products, system dynamics models for kinematics and vibration analysis to modeling and analysis of civil infrastructures, network models for communication and transportation systems, enterprise and supply chain models, and simulation and gaming models for training, situation assessment, and education.

In Chapter 8, George Karniadakis discusses validation, verification and uncertainty quantification (UQ) for SBE&S. Despite its importance throughout SBE&S, the field of UQ is relatively underdeveloped. However, there is intense renewed interest in modeling and quantifying uncertainty, and in verification and validation (V&V) of large-scale simulations. While *verification* is the process of determining that a model implementation accurately represents the developer's conceptual description of the model and the solution to the model, *validation* is the process of determining the degree to which a model is an accurate representation of the real world from the perspective of the intended uses of the model. Rapid advances in SBE&S capabilities that will allow larger and more complex simulations than ever before urgently require new approaches to V&V and UQ.

Peter Cummings discusses multiscale methods for SBE&S in Chapter 9. Multiscale methods research involves the development and application of techniques that allow modeling and simulation of phenomena across disparate time and length scales that, in many cases, span many orders of magnitude. True multiscale simulations, with both automated upscaling and adaptive downscaling, are still rare because there has been no broadly successful general multiscale modeling simulation methodology developed to date. The need for such methodology, however, continues to grow, and holds the key to making SBE&S more broadly applicable in areas such as design, control, optimization, and abnormal situation modeling.

Sangtae Kim describes issues and opportunities for data-intensive applications of SBE&S, including visualization, in Chapter 10. The life sciences and particle physics communities are pushing the envelope of data-intensive science. As computational capabilities continue to grow, big data, visualization and dynamic data-driven simulations are crucial technology elements in the solution of problems ranging from global climate change to earthquake prediction to the production of transportation fuels from the last remaining oil fields. However, insufficient cyberinfrastructure investments at university campuses particularly in the United States are hampering progress.

Finally, in Chapter 11 Celeste Sagui discusses the urgent global need for educating and training the next generation of researchers and innovators in SBE&S. From university campuses to industrial employers and government research facilities, the difficulty in finding talent with breadth and depth in their discipline as well as competency in scientific computation was the number one issue for concern identified by this panel. New paradigms for computational science education are emerging and are highlighted in this chapter.

Appendix A contains biographies of the panelists and advisors. Trip reports for each of the sites visited in Asia and Europe during the international assessment are available on the International Assessment of Research and Development in Simulation-Based Engineering and Science website, http://www.wtec.org/private/sbes/. The required password is available from WTEC upon request. The survey questions for host organizations are listed in Appendix B. A bibliometric analysis of simulation research is presented in Appendix C. Finally, a glossary of terms is provided in Appendix D. Additional information and documentation for all phases of the WTEC International Assessment of R&D in Simulation-Based Engineering and Science are available on the WTEC website at http://www.wtec.org/sbes.

References

Atkins, D. M. (2006). *NSF's Cyberinfrastructure Vision for 21st Century Discovery*, version 7.1 (National Science Foundation, USA), http://nsf.gov.

Benioff, M. R. and Lazowska, E. D. (2005). *Computational Science: Ensuring America's Competitiveness* (President's Information Technology Advisory Committee, USA), http://www.nitrd.gov/pitac/reports/20050609_computational/computational.pdf.

Bramley, R., Char, B., Gannon, D., Hewett, T., Johnson, C. and Rice, J. (2000). Enabling technologies for computational science: Frameworks, middleware and environments, *Proc. Wkp. on Sci. Knowledge, Information, and Computing*, eds. Houstis, E., Rice, J., Gallopoulos, E. and Bramley, R. (Kluwer), pp. 19–32.

Challenges in High-End Computing (2007). http://www.epsrc.ac.uk/CMSWeb/Downloads/Other/ChallengesHEC.pdf.

Committee on Science, Engineering, and Public Policy and Committee on Prospering in the Global Economy of the 21st Century (2007). *Rising Above the Gathering Storm: Energizing and Employing America for a Brighter Economic Future.* (National Academies Press, USA), http://www.nap.edu/catalog.php?record_id=11463.

Council on Competitiveness (2004). *Final Report from the High Performance Users Conference: Supercharging U.S. Innovation and Competitiveness* (Council on Competitiveness, USA).

Coveney, P. *et al.* (2006). *U.K. Strategic Framework for High-End Computing*, http://www.epsrc.ac.uk/CMSWeb/Downloads/Other/2006HECStrategicFramework.pdf.

Department of Defense (2002). Report on High Performance Computing for the National Security Community (DoD, USA), http://WWW.HPCMO.HPC.mil/Htdocs/DOCUMENTS/04172003_hpc_report_unclass.pdf.

Department of Defense, Defense Science Board (2000). *Report of the Defense Science Board Task Force on DoD Supercomputing Needs* (DoD, USA), http://www.acq.osd.mil/dsb/reports/dodsupercomp.pdf.

Department of Energy, Office of Science (2000). *Scientific Discovery Through Advanced Computing* (DoE, USA).

European Science Foundation, European Computational Science Forum (2007). *The Forward Look Initiative. European Computational Science: The Lincei Initiative: From Computers to Scientific Excellence* (ESF, France), http://www.esf.org/activities/forward-looks/all-current-and-completed-forward-looks.html.

European Strategy Forum for Research Infrastructures (2006). *European Roadmap for Research Infrastructures: Report 2006* (Office for Official Publications of the European Communities, Luxembourg), ftp://ftp.cordis.europa.eu/pub/esfri/docs/esfri-roadmap-report-26092006_en.pdf.

Executive Office of the President, Office of Science and Technology Policy (2004). Federal Plan for High-End Computing: Report of the High-End Computing Revitalization Task Force (OSTP, USA).

Federal High Performance Computing Program (1993). *Wkp. and Conf. on Grand Challenges Applications and Software Technology* (Federal High Performance Computing Program, USA).

Interagency Panel on Large Scale Computing in Science and Engineering (1982). *Report of the Panel on Large Scale Computing in Science and Engineering* (DoD and NSF, USA).

President's Information Technology Advisory Committee (1999), *Report to the President: Information Technology Research: Investing in Our Future*, eds. Joy, W. and Kennedy, K. (PITAC, USA), http://www.nitrd.gov/pitac/report/.

Joseph, E., Snell, A. and Willard, C. G. (2004). *Council on Competitiveness Study of U.S. HPC Users* (Council on Competitiveness, USA), http://www.compete.org/pdf/HPC_Users_Survey.pdf.

Kaufmann, N. J., Willard, C. G., Joseph, E. and Goldfarb, D. S. (2003). *Worldwide High Performance Systems Technical Computing Census* (IDC, USA).

Keyes, D., Colella, P., Dunning, T. H., Jr. and Gropp, W. D. (2003). *A Science-Based Case for Large-Scale Simulation*, vol. 1, (DOE, USA), http://www.pnl.gov/scales/.

Keyes, D., Colella, P., Dunning, T. H., Jr. and Gropp, W. D. (2004). *A Science-Based Case for Large-Scale Simulation*, vol. 2, (DOE, USA), http://www.pnl.gov/scales/.

National Research Council, Computer Science and Telecommunications Board (2005). *Getting Up to Speed: the Future of Supercomputing* (National Academies Press, USA).

National Research Council (2008a). *National Academy of Engineering Report of Committee on Integrated Computational Materials Engineering* (National Academies Press, USA).

National Research Council (2008b). *The Potential Impact of High-End Capability Computing on Four Illustrative Fields of Science and Engineering* (National Academies Press, USA).

National Science Foundation, Blue Ribbon Panel on High Performance Computing (1993). *From Desktop to Teraflop: Exploiting the U.S. Lead in High Performance Computing* (NSF, USA).

National Science Foundation (1995). *Report of the Task Force on the Future of the NSF Supercomputer Centers Program* (NSF, USA).

National Science Foundation (2003). *Revolutionizing Science and Engineering through Cyberinfrastructure: Report of the National Science Foundation Blue Ribbon Advisory Panel on Cyberinfrastructure* (NSF, USA).

Oden, J. T., Belytschko, T., Hughes, T. J. R., Johnson, C., Keyes, D., Laub, A., Petzold, L., Srolovitz, D. and Yip, S. (2006). *Revolutionizing Engineering Science through Simulation: A Report of the National Science Foundation Blue Ribbon Panel on Simulation-Based Engineering Science* (NSF, USA), http://www.nsf.gov/pubs/reports/sbes_final_report.pdf.

President's Council of Advisors on Science and Technology (2004). *Report to the President: Sustaining the Nation's Innovation Ecosystems, Information Technology, Manufacturing, and Competitiveness* (PCAST, USA).

Society for Industrial and Applied Mathematics, Working Group on Computational Science and Engineering Education (2001). Graduate education in computational science and engineering, *SIAM Review*, 43(1), 163–177.

Chapter 2

LIFE SCIENCES AND MEDICINE

Linda Petzold

2.1 Introduction

There is an unprecedented opportunity in the next few decades to make major advances in the life sciences and medicine through simulation. The mapping of the genome, together with high-throughput experimental techniques, imaging techniques, and the power of today's algorithms and computers, have the imminent potential to revolutionize the biological and medical sciences.

Some of the most important questions in the fields of biology, chemistry, and medicine remain unsolved as a result of our limited understanding of the structure, behavior, and interaction of biologically significant molecules. Many of these questions could in principle be answered if it were possible to perform detailed atomic-level molecular dynamics (MD) simulations of sufficient accuracy and duration. Important biological phenomena that might otherwise be explored using such simulations, however, occur over time scales on the order of a millisecond — about three orders of magnitude beyond the reach of the longest current MD simulations. Moreover, it remains unclear, because current simulations are not long enough to tell, whether the molecular mechanics force fields on which such simulations are typically based would be sufficiently accurate to fully elucidate these phenomena or if more complex force fields would need to be employed.

Modeling and simulation have been in use for some time in the pharmaceutical industry. Initial applications have focused mainly on chemistry and molecular structure. However, new techniques for modeling variability in patient physiology, as well as new software platforms, lower-cost computational resources, and an explosion of available data, have fueled recent advances in "predictive biosimulation" — the dynamic simulation of biological systems. Predictive biosimulation has been used to support the development of new therapeutic drugs, meeting the pharmaceutical industry's need for new approaches to deal with the complexity of disease and the increased risk and expense of research and development.

Several factors are likely to accelerate adoption of biosimulation technology in the industry. The FDA, through its Critical Path Initiative, has recognized the importance of modeling and simulation and is starting to commit more resources toward making biosimulation part of the evaluation process for new therapies. Academic groups and companies are increasingly specializing in the field and building active collaborations with industry. Pharmaceutical companies are creating and expanding internal systems biology groups that incorporate biological modeling. The potential impacts for systems biology in the pharmaceutical industry are huge. An average of $1.4 billion and 14 years are spent on the development of a successful drug (Bangs, 2007). The failure rate for new drugs is greater than 90%. Meanwhile, annual spending on new drugs increases year after year, while drug approvals are decreasing. Modeling and simulation can assist in drug development through identifying targets of intervention; determining the relationships for dose, exposure, and response; and assessing development strategies and trial designs in populations. Using a population of virtual patients, clinical variability can be explored and risk analysis undertaken. Another big variability amenable to modeling and simulation is the heterogeneity of causes and mechanisms of nearly all clinical diagnoses. The groups of patients who might benefit from a drug or who could be at risk may be able to be identified through these technologies.

Modeling of whole organs and complete physiological systems is becoming possible. Modeling of complete cardiovascular systems provides information and insight both for medical research and for the evaluation of surgical options. Modeling provides information both for cancer research and for laser surgery. Neuromuscular biomechanics seeks fundamental understanding of the mechanisms involved in the production of movement and is motivated by opportunities to improve treatments for cerebral palsy. Biomechanical simulation is used in surgical planning and design of physical therapies. Rapidly increasing amounts of data, combined with high-performance algorithms and computers, have made modeling the brain in full experimental detail a realistic goal for the foreseeable future.

Where are the computational bottlenecks? Many of the models are multiscale, requiring new algorithms. Of critical importance are the need to deal with potentially large amounts of heterogeneous data and the need to communicate the source, the uncertainties, and the conditions under which the data were obtained. As models become more and more complex, they will require high-performance computing. This work is highly multidisciplinary — it requires a team of researchers, each of whom is expert in one discipline but knowledgeable in others. U.S. graduate schools are currently producing relatively few such PhD graduates. In the following, we examine the status of the areas of molecular dynamics, systems biology, and biophysical modeling in more detail.

2.2 Molecular Dynamics

Molecular dynamics simulations can be used to model the motions of molecular systems, including proteins, cell membranes, and DNA, at an atomic level of detail. A sufficiently long and accurate MD simulation could allow scientists and drug designers to visualize for the first time many critically important biochemical phenomena that cannot currently be observed in laboratory experiments, including the folding of proteins into their native three-dimensional structures, the structural changes that underlie protein function, and the interactions between two proteins or between a protein and a candidate drug molecule (Brooks and Case, 1993; Frenkel and Smit, 2001; Karplus and McCammon, 2002; Schlick *et al.*, 1999). Such simulations could in principle answer some of the most important open questions in the fields of biology and chemistry, and they have the potential to make substantial contributions to the process of drug development (Shaw *et al.*, 2007).

A classical MD computation simulates the motion of a collection of atoms over a period of time according to the laws of Newtonian physics. The chemical system might consist of a protein and its surrounding solvent environment (which typically includes water and various ions), and might include other types of molecules such as lipids, carbohydrates, nucleic acids, or drug molecules. The entire chemical system occupies a small volume, typically tens of angstroms on a side, filled with tens or hundreds of thousands of atoms (Adcock and McCammon, 2006). The computation strains the resources of even the largest, most advanced computers; the vast majority of such MD simulations have been less than a microsecond in length, whereas many of the most important biological processes, such as protein folding, interactions between proteins, binding of drugs to their molecular targets, and the dynamics of conformational changes, often occur over time scales on the order of a millisecond.

To accomplish a millisecond simulation will require the use of a massive number of arithmetic processing units. Because of the global nature of the inter-atomic force calculations, this will necessitate a great deal of inter-processor communication, which must be very carefully managed in order to maintain efficiency. Further complicating the path to the millisecond timescale is the fact that current simulations are not long enough to tell whether commonly used force field approximations will be sufficiently accurate on longer timescales. Some known deficiencies of the most commonly used force fields include the omission of electrostatic polarization effects, the lack of explicitly represented hydrogen bonds, the use of simplified harmonic and trigonometric models to model the energetics of covalent bonds, and nonphysical repulsive Van der Waals terms. Some steps are already being taken to improve these approximations, but there may ultimately be a need to develop new biophysical theory.

There has been much recent progress in this area, and there is a huge opportunity for impact when simulation technologies reach the millisecond timescale. Current MD codes include AMBER (Case *et al.*, 2005), CHARMM (Brooks *et al.*, 1983), GROMACS (University of Groningen, 1993, The Netherlands, generally regarded as the fastest commonly used single-processor code) and the parallel MD codes NAMD (University of Illinois, Urbana-Champaign, 1996, USA), LAMMPS (Sandia National Labs, 1999, USA), Blue Matter (IBM Research, 2002, USA), and Desmond (D. E. Shaw Research, 2006, USA). The greatest performance gains may ultimately be achieved using special-purpose MD machines, which have included the Delft MD Processor (Technische Universiteit Delft, 1988, The Netherlands), FASTRUN (Columbia University, 1990, USA), MD Engine (Fuji-Xerox, 1997, Japan), MD-GRAPE (University of Tokyo/RIKEN, 1996, Japan), and Anton (D. E. Shaw Research, 2008, USA). The recent arrival of GPGPU (general purpose graphics processing units) computing is likely to revolutionize MD simulation because the underlying algorithms in MD require fine-grained parallelism, which is ideally suited to these architectures. NAMD may be the first open-source MD code in use by the biosimulation community to exploit GPGPUs, as the code is currently being ported to NVIDIA processors using CUDA.

The United States has historically played a leadership role in a number of research areas that have driven the development of new algorithms, architectures, and applications for MD simulations. However, this role is now endangered by funding limitations at the National Institutes of Health (NIH), the National Science Foundation (NSF), the Defense Advanced

Research Projects Agency (DARPA), and other federal agencies, and by the loss of PhD students and faculty to other disciplines and to the private sector.

2.3 Systems Biology

Systems biology has been defined by a previous WTEC study as "the understanding of network behavior, and in particular their dynamic aspects, which requires the utilization of mathematical modeling tightly linked to experiment" (Cassman *et al.*, 2005, p. 1). Systems biology bridges molecular biology and physiology. At the core of this field is the focus on networks, where the goal is to understand the operation of systems, as contrasted with the reductionist tradition in biology of understanding the component parts. This involves a variety of approaches, including the identification and validation of networks, the creation of appropriate datasets, the development of tools for data acquisition and software development, and the use of modeling and simulation software in close linkage with experiment, often done to understand dynamic processes. The potential for systems biology is huge, both for advancing the understanding of fundamental biological processes, and for enabling predictive simulation in pharmaceutical research and development. Systems biology is particularly relevant to research in cancer and in diabetes, where it offers the potential of identifying targets from novel pathways. The WTEC panelists visited several institutions that are moving the systems biology field forward in significant ways.

2.3.1 *Systems Biology Institute, Japan*

One of the world's leading groups in systems biology is the Systems Biology Institute (SBI) in the Department of Systems Biology at the Japan Foundation for Cancer Research in Tokyo (see Asia Site Reports on the International Assessment of Research and Development in Simulation-Based Engineering and Science website, http://www.wtec.org/private/sbes/. The required password is available from WTEC upon request.). This group is led by Dr. Hiroaki Kitano, who is often referred to as the "father of systems biology." This group is well known for its software as well as for its research results. Its current focus is on the development of experimental data and software infrastructure. The software infrastructure includes Systems Biology Markup Language (SBML), Systems Biology

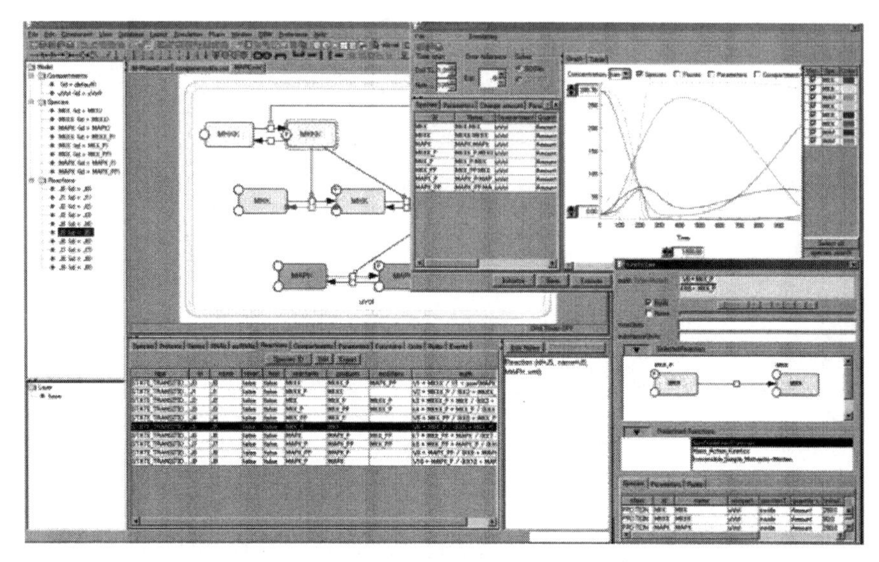

Fig. 2.1. CellDesigner is a structured diagram editor for drawing gene-regulatory and biochemical networks. Networks are drawn based on the process diagram, with a graphical notation system proposed by Kitano, and are stored using the Systems Biology Markup Language (SBML), a standard for representing models of biochemical and gene-regulatory networks. Networks are able to link with simulation and other analysis packages through Systems Biology Workbench (SBW) (http://www.systems-biology.org/cd/).

Graphical Notation (SBGN) (Kitano *et al.*, 2005), CellDesigner (Funahashi *et al.*, 2003), and Web 2.0 Biology, designed for the systematic accumulation of biological knowledge (see Fig. 2.1). Development of systems biology as a field requires an extensive software infrastructure. Recognizing that it is difficult to publish software, the merit system in this lab values software contributions as well as publications. Biological systems under investigation include cancer robustness, type-2 diabetes, immunology, infectious diseases, metabolic oscillation, cell cycle robustness, and signaling network analysis. The experimental infrastructure under development includes the gTOW assay described below, microfluidics, and tracking microscopy.

Key issues in cellular architecture that are being investigated at the SBI include the implications of robustness versus fragility on cellular architecture. Some systems have to be unstable to be robust (for example, cancer). Points of fragility in a system are important to the determination of drug effectiveness. SBI researchers use the robustness profile to reveal principles of cellular robustness, to refine computer models, and to find therapeutic targets. They use yeast (budding yeast and fission yeast) as the model organism and cell cycle as the model system. The models are currently ODE (ordinary differential equation) models. The limits of parameters are used as indicators of robustness; for example, how much can you increase or decrease parameters without disrupting the cell cycle? *This is a different type of uncertainty analysis than most of the current work in the United States.*

Experimental methods are required that can comprehensively and quantitatively measure parameter limits. The SBI has developed Genetic Tug of War (gTOW), an experimental method to measure cellular robustness. gTOW introduces extra copies of each gene to see how much it changes the cell cycle. There are implications of this technology for drug development.

2.3.2 *Vrije University, The Netherlands*

The theme of the systems biology center BioCentrum Amsterdam at Vrije Universiteit, led by Professor Hans Westerhoff, another worldwide leader in systems biology, is "to cure a disease, one must cure the network." Professor Westerhoff heads a transnational research group on Systems Biology that includes the Manchester Centre for Integrative Systems Biology (MCISB) in the Manchester Interdisciplinary BioCentre (MIB) and the BioCentrum Amsterdam. He is also the Director of the University of Manchester Doctoral Training Centre in Integrative Systems Biology (supported by the UK's Biotechnology and Biological Sciences Research Council [BBSRC] and the Engineering and Physical Sciences Research Council [EPSRC]). Here, a focus on the development of software infrastructure is also apparent.

BioCentrum Amsterdam is a driver behind the Silicon Cell program (Westerhoff, 2001; http://www.siliconcell.net; see also Europe Site Reports on the International Assessment of Research and Development in Simulation-Based Engineering and Science website, http://www.wtec.org/private/sbes/. The required password is available from WTEC upon request). This is an initiative to construct computer replicas of parts of living cells from detailed experimental information. The Silicon Cells can then be used in analyzing functional consequences that are not directly evident from component data but arise through contextual dependent interactions. Another prominent effort at the center is in network-based drug design. The premise is that the recent decline in drug discovery may be due to the focus on single molecules, neglecting cellular, organellar, and whole-body responses. A specific research effort focuses on the treatment of African sleeping sickness using trypanosome glycolysis as a drug target. This has entailed development of an ODE model of trypanosome glycolysis and sensitivity analysis to identify those reactions that can best control the flux. Using the model, it was found that the network behavior was complicated by the fact that gene-expression response can either counteract or potentiate the primary inhibition.

2.3.3 *Technical University of Denmark*

The Center for Biological Sequence Analysis (CBS) at the Technical University of Denmark (see Europe Site Reports on the International Assessment of Research and Development in Simulation-Based Engineering and Science website, http://www.wtec.org/private/sbes/. The required password is available from WTEC upon request) is home to another world-class effort in systems biology. Over the last decade, the CBS has produced a large number of bioinformatics tools. The codes are very popular, as evidenced by the number of page-views to the CBS webpages (over 2 million a month). One of the major challenges in this field is the integration of data. The amount of data generated by biology is exploding. CBS has a relational data warehouse comprising 350+ different databases. In order to handle data integration of 120 terabyte size, CBS has developed its own integration tool. Ultimately, CBS researchers aim to develop data integration systems that can talk to one another. The implications for both scientific research and its culture are clearly expressed in the December 2005 *Nature* editorial "Let data speak to data": "Web tools now allow data sharing and informal debate to take place alongside published papers. But to take full advantage, scientists must embrace a culture of sharing and rethink their vision of databases."

The CBS group is moving towards other frontiers, the "disease interactomes." This involves the use of text mining to relate proteins to diseases, or the identification of new disease gene candidates by the phenomic ranking of protein complexes (Lage *et al.*, 2007). This new research leads to new types of "biobanks" with new computational challenges: (1) finding disease genes, and their "systemic" properties, in cases where the environment also plays a major role, and (2) extracting information from complex, messy "databases" and registries across countries. The combining of medical informatics with bioinformatics and systems biology represents a new trend in disease gene finding and phenotype association, which can also include social and behavioral levels. Linking medical informatics with bioinformatics and systems biology requires bridging the gap between the molecular level and the phenotypic clinical levels, and linking two exponentially growing types of computer-accessible data: biomolecular databases and their clinical counterparts. This necessitates development of a systems biology that recognizes a changing environment.

2.3.4 *U.S. Systems Biology Efforts*

Major efforts in systems biology exist also in the United States, with comparable impacts. Systems Biology Markup Language (Hucka *et al.*, 2004), a key development that enables the sharing of models, was developed jointly by the SBI and the California Institute of Technology (Caltech). The Complex Pathway Simulator (COPASI) (Hoops *et al.*, 2006), a software application for simulation and analysis of biochemical networks, is a collaborative project led by P. Mendes of Virginia Polytechnic Institute (Virginia Tech) and the University of Manchester (UK), and U. Kummer of Heidelberg University (Germany). Major modeling and experimental efforts exist at the Massachusetts Institute of Technology (MIT), the Institute for Systems Biology in Seattle, Caltech, the University of Washington, and the University of California's Berkeley, San Diego, and Santa Barbara campuses, among other U.S. universities. BioSPICE, an initiative that DARPA began in 2001 and ran for five years (Kumar and Feidler, 2003; http://biospice.lbl.gov/home.html), was the impetus behind a tremendous surge in systems biology research and software development in the United States.

2.4 Biophysical Modeling

Traditionally, physiology has been concerned with the integrative function of cells, organs, and whole organisms. As more detailed data has become available at the molecular level, it has become more difficult to relate whole organ function to the underlying detailed mechanisms that exploit this molecular knowledge. Organ and whole organism behavior needs to be understood at both a systems level and in terms of subcellular function and tissue properties.

Linking the vast amounts of information ranging from the genome to medical imaging poses many challenges for SBE&S. It requires the development of databases of information at all spatial and temporal scales (see Fig. 2.2), the development of models and algorithms at a wide range of

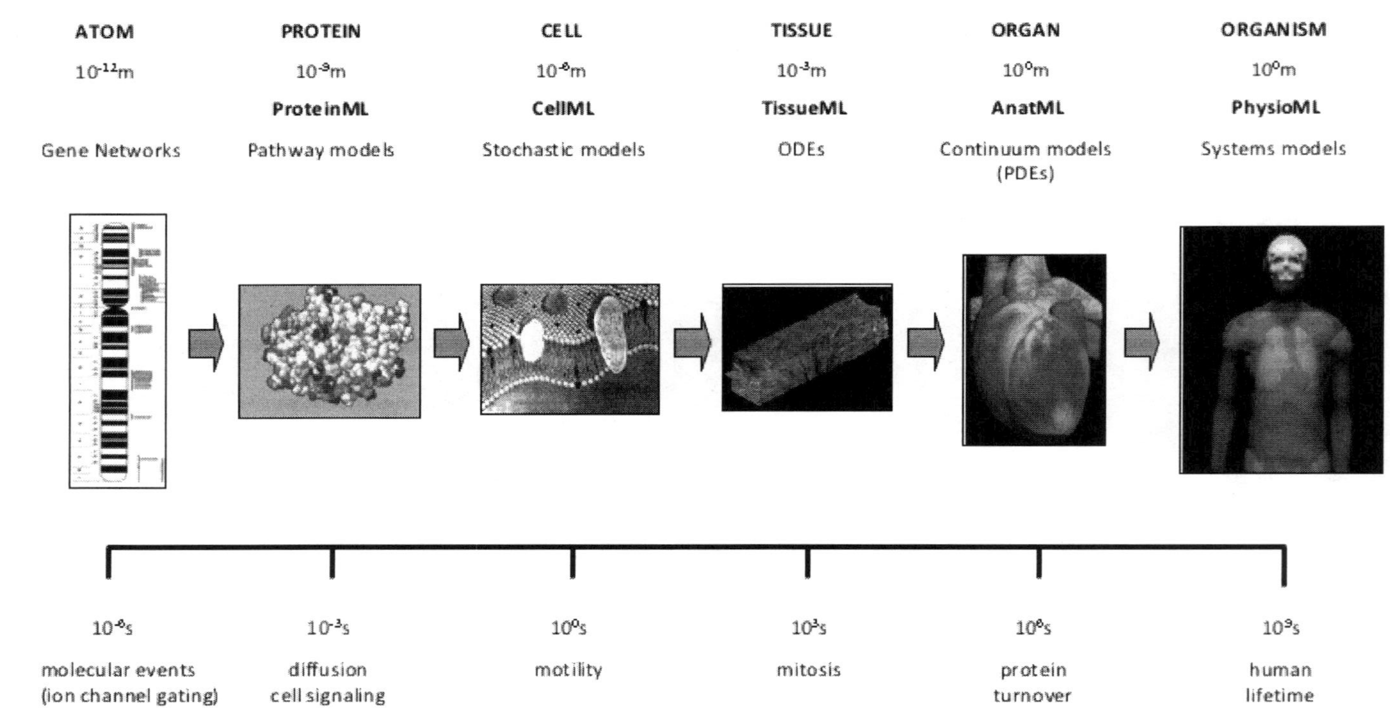

Fig. 2.2. Spatial (top) and temporal (bottom) scales encompassed by the Human Physiome Project. Markup languages (PhysioML, AnatML, TissueML, CellML) are defined for each spatial level. The types of mathematical model appropriate to each spatial scale are also indicated. (Hunter *et al.*, 2002; figure courtesy of P. Hunter).

scales, the mathematical and software technology for linking the models so that fine-scale information is accessible by coarser models, and the effective use of petascale high-performance computers.

2.4.1 *International Physiome Project*

A great deal of work in this area has been taking place under the auspices of the Physiome Project (see http://en.wikipedia.org/wiki/Physiome). The Physiome Project is a worldwide effort focused on compiling laboratories into a single, self-consistent framework. Research initiatives related to Physiome include the Wellcome Trust Physiome Project, the IUPS (International Union of Physiological Sciences) Physiome Project, the EuroPhysiome Project, and the Physiome project of the NSR (National Simulation Resource) at the University of Washington Department of Bioengineering (Hunter *et al.*, 2002). The Life Science core focus of the Riken next-generation supercomputer center in Japan (see Asia Site Reports on the International Assessment of Research and Development in Simulation-Based Engineering and Science website, http://www.wtec.org/private/sbes/. The required password is available from WTEC upon request.) center on the multiscale modeling of the virtual human, from genes to proteins, to cells, to tissues and organs.

In groundbreaking work associated with Physiome, a mathematical model of the heart has been developed (Smith *et al.*, 2001) that encompasses the anatomy and cell and tissue properties of the heart and is capable of revealing the integrated physiological function of the electrical activation, mechanics, and metabolism of the heart under a variety of normal and pathological conditions. This model has been developed over the last 20 years as a collaborative effort by research groups at the University of Auckland, the University of Oxford, and the University of California, San Diego. A model of the lungs, including gas transport and exchange, pulmonary blood flow, and soft tissue mechanics, is being undertaken as a collaboration between universities in the UK, the United States, and New Zealand (Howatson *et al.*, 2000).

2.4.2 *EPFL Arterial Map*

At Ecole Polytechnique Fédérale de Lausanne (EPFL), there is a project led by Professor A. Quarteroni for extracting geometry from raw data to get a 3D "map" of the arterial system (see Europe Site Reports on the International Assessment of Research and Development in Simulation-Based Engineering and Science website, http://www.wtec.org/private/sbes/. The required password is available from WTEC upon request). These reconstructions are then turned into a finite element structure that provides the basis for detailed simulations of blood flow. Particular attention has been paid to difficulties in attaching different length and time scales (the multiscale problem) to the cardiovascular system.

2.4.3 *EPFL Blue Brain Project*

Perhaps the ultimate frontier in biophysical modeling is to understand, in complete experimental detail, the structure and function of the brain. An impressive effort is underway in this area at the EPFL, in the Blue Brain project directed by neuroscientist Henry Markram. Named after the IBM Blue Gene supercomputer it relies on, the Blue Brain project (see Europe Site Reports on the International Assessment of Research and Development in Simulation-Based Engineering and Science website, http://www.wtec.org/private/sbes/. The required password is available from WTEC upon request.) has the goal of modeling, in every detail, the cellular infrastructure and electrophysiological interactions within the cerebral neocortex, which represents about 80% of the brain and is believed to house cognitive functions such as language and conscious thought. The project started out to reverse-engineer the neocortex. This involved collecting experimental data for different types of cells, types of electrical behavior, and types of connectivity. The model includes 10,000 neurons, 340 types of neurons, 200 types of ion channels, and 30 million connections.

EPFL's Blue Brain model is a detailed multiscale model, related to variables as measured in a laboratory. It is an enormous system of ordinary differential equations, modeling electrochemical synapses and ion channels. The main objective of the model is to capture electrical behavior. The model has been implemented in such a way that it is easy to modify and add to it. The current capability is a faithful *in silico* replica at the cellular level of a neocortical column of a young rat (Fig. 2.3). A goal is to capture and explain diversity observed in the lab: electrical, morphological, synaptic, and plasticity. Building a database of experimental results is a major part of this effort. Some of the long-term objectives are to model a normal brain and a diseased brain to identify what is going wrong. Another objective is to evaluate emergent network phenomena. A demonstration was given in which the model replicated slice experiments that exhibit

Fig. 2.3. A visualization of the neocortical column model with 10,000 neurons. In false colors, the membrane voltage for all simulated compartments is shown, indicating the activity of the network. Image © 2009 by BBP/EPFL.

network-scale phenomena. In this way, it may someday be possible to do *in silico* pharmacology experiments.

The Blue Brain Project models 10,000 cells involving approximately 400 compartments per neuron with a cable-equation, Hodgkin–Huxley model. There are 30 million dynamic synapses. The computing platform uses a 4 rack IBM BlueGene/L with a 22.4 Tflop peak and 8192 processors. Additionaly, an SGI Prism Extreme is used for visualization. The project features an impressive integration of experiment, modeling, high-performance computing, and visualization.

2.4.4 *U.S. Biophysical Modeling Efforts*

In the United States, biophysical modeling is being undertaken at the Stanford University Simbios Center for Vascular Biomechanics Research (Taylor and Draney 2004; http://simbios.stanford.edu/index.html), and at Brown University. Grinberg *et al.* (2008) review the issues involved in order to make detailed 3D simulations of blood flow in the human arterial tree feasible. A straightforward approach to the detailed computation is computationally prohibitive even on petaflop computers, so a three-level hierarchical approach based on the vessel size is utilized, consisting of a Macrovascular Network (MaN), a Mesovascular Network (MeN), and a Microvascular Network (MiN). A multiscale simulation coupling MaN–MeN–MiN and running on hundreds of thousands of processors on petaflop computers is projected to be possible and will require no more than a few CPU hours per cardiac cycle, hence opening up the possibility of simulation studies of cardiovascular diseases, drug delivery, perfusion in the brain, and other pathologies.

2.5 Summary of Key Findings

The role of SBE&S in life sciences and medicine is growing very rapidly and is critical to progress in many of these areas. For SBE&S to realize its potential impacts in life sciences and medicine will require:

- Large, focused multidisciplinary teams. Much of the research in this area requires iteration between modeling and experiment.
- Integrated, community-wide software infrastructure for dealing with massive amounts of data, and addressing issues of data provenance, heterogeneous data, analysis of data, and network inference from data.
- High-performance algorithms for multiscale simulation on a very large range of scales and complexities.
- High-performance computing and scalable algorithms for multicore architectures. In particular, petascale computing can finally enable molecular dynamics simulation of macromolecules on the biologically interesting millisecond timescale.
- Techniques for sensitivity and robustness analysis, uncertainty analysis, and model invalidation.
- Visualization techniques, applicable to massive amounts of data, that can illustrate and uncover relationships between data.
- Appropriately trained students who are conversant in both the life sciences and SBE&S and can work in multidisciplinary teams. This was mentioned by hosts at virtually all the WTEC panelists site visits as the biggest bottleneck to progress in the field.
  What is the status of SBE&S capabilities in life sciences and medicine worldwide?
- WTEC panelists are aware of world-class research in SBE&S in molecular dynamics, systems biology, and biophysical modeling in the United States, Europe, and Japan, and in some specific areas, for example, on increasing the potency of natural medicines in China.

- The quality of leading researchers in the United States is comparable to that of leading researchers in Europe and Japan.
- Infrastructure (access to computing resources and software professionals) and funding models to support ambitious, visionary, long-term research projects are much better in Europe and Japan than in the United States.
- Funding models and infrastructure to support multi-investigator collaborations between academia and industry are much more developed in Europe than in the United States.
- Support for the development of community software is much stronger in Europe and Japan than in the United States in recent years, following the decline in DARPA funding in this area.

References

Adcock, S. A. and McCammon, J. A. (2006). Molecular dynamics: survey of methods for simulating the activity of proteins, *Chem. Rev.*, 106, 1589–1615.

Bangs, A. (2007). Predictive biosimulation in pharmaceutical R&D, *WTEC Wkp on Rev. of Simulation-Based Eng. and Sci. Rsch. in the United States*, Arlington, VA, 1–2 November.

Brooks, B. R., Bruccoleri, R., Olafson, B., States, D., Swaminathan, S. and Karplus, M. (1983). CHARMM: a program for macromolecular energy, minimization and dynamics calculations, *J. Comp. Chem.*, 4, 187–217.

Brooks, C. I. and Case, D. A. (1993). Simulations of peptide conformational dynamics and thermodynamics, *Chem. Rev.*, 93, 2487–2502.

Case, D., Cheatham, T., III, Darden, T., Gohlke, H., Luo, R., Merz, K., Jr., Onufriev, A., Simmerling, C., Wang, B. and Woods, R. (2005). The Amber biomolecular simulation programs, *J. Comp. Chem.*, 26, 1668–1688.

Cassman, M., Arkin, A., Doyle, F., Katagiri, F., Lauffenburger, D. and Stokes, C. (2005). *WTEC Panel Report on International Research and Development in Systems Biology* (WTEC, Baltimore).

Frenkel, D. and Smit, B. (2001). *Understanding Molecular Simulation; From Algorithms to Applications*, second ed. (Academic Press, London).

Funahashi, A., Morohashi, M., Kitano, H. and Tanimura, N. (2003). CellDesigner: a process diagram editor for gene-regulatory and biochemical networks, *Biosilico.*, 1(5), 159–162.

Grinberg, L., Anor, T., Madsen, J., Yakhot, A. and Karniadakis, G. E. (2008). Large-scale simulation of the human arterial tree, *Clin. Exper. Pharmacol. Physiol.*, to appear.

Hoops, S., Sahle, S., Gauges, R., Lee, C., Pahle, J., Simus, N., Singhal, M., Xu, L., Mendes, P. and Kummer, U. (2006). COPASI: a complex pathway simulator, *Bioinformatics*, 22, 3067–3074.

Howatson, M., Pullan, A. J. and Hunter, P. J. (2000). Generation of an anatomically based three-dimensional model of the conducting airways, *Ann. Biomed. Eng.*, 28, 793–802.

Hucka, M., Finney, A., Bornstein, B. J., Keating, S. M., Shapiro, B. E., Matthews, J., Kovitz, B. L., Schilstra, M. J., Funahashi, A., Doyle, J. C. and Kitano, H. (2004). Evolving a *lingua franca* and associated software infrastructure for computational systems biology: the Systems Biology Markup Language (SBML) project, *Syst. Biol.*, 1(1), 41–53.

Hunter, P., Robbins, P. and Noble, D. (2002). The IUPS human physiome project, *Eur. J. Physiol.*, 445, 1–9.

Karplus, M. and McCammon, J. A. (2002). Molecular dynamics simulations of biomolecules, *Nat. Struct. Biol.*, 9, 646–652.

Kitano, H., Funahashi, A., Matsuoka, Y. and Oda, K. (2005). Using process diagrams for the graphical representation of biological networks, *Nat. Biotechnol.*, 23, 961–966.

Kumar, S. P. and Feidler, J. C. (2003). BioSPICE: a computational infrastructure for integrative biology, *OMICS: A J. Integrative Biol.*, 7(3), 225.

Lage, K., Karlberg, E. O., Storling, Z. M., Olason, P. I., Pedersen, A. G., Rigina, O., Hinsby, A. M., Tumer, Z., Pociot, F., Tommerup, N., Moreau, Y. and Brunak, S. (2007). A human phenome-interactome network of protein complexes implicated in genetic disorders, *Nat. Biotechnol.*, 25, 309–316.

Lehrer, J. (2008). Out of the blue, *Seed Magazine*, March 3, http://www.seedmagazine.com/news/2008/03/out_of_the_blue.php.

Nagar, B., Bornmann, W., Pellicena, P., Schindler, T., Veach, D., Miller, W. T., Cloarkson, B. and Kuriyan, J. (2002). Crystal structures of the kinase domain of c-Abl in complex with the small molecular inhibitors PD173955 and Imatinib (STI-571), *Cancer Res.*, 62, 4236–4243.

Nature editors (2005). Let data speak to data, *Nature*, 438, 531.

Schlick, T., Skeel, R. D., Brunger, A. T., Kale, I. V., Board, J. A., Hermans, J. and Schulte, K. (1999). Algorithmic challenges in computational molecular biophysics, *J. Comput. Phys.*, 151(1), 9–48.

Shaw, D. E., Deneroff, M., Dror, R., Kuskin, J., Larson, R., Salmon, J., Young, C., Batson, B., Bowers, K., Chao, J., Eastwood, M., Gagliardo, J., Grossman, J., Ho, C. R., Ierardi, D., Kolossvary, I., Klepeis, J., Layman, T., McLeavey, C., Moraes, M., Mueller, R., Priest, E., Shan, Y., Spengler, J., Theobald, M., Towles, B. and Wang, S. (2007). Anton, a special-purpose machine for molecular dynamics simulation, *Proc. 34th Ann. Intl. Symp. Comp. Arch.*, ACM, pp. 1–12.

Smith, N. P., Mulquiney, P. J., Nash, M. P., Bradley, C. P., Nickerson, D. P. and Hunter, P. J. (2001). Mathematical modeling of the heart: cell to organ, *Chaos, Solitons Fractals*, 13, 1613–1621.

Taylor, C. A. and Draney, M. T. (2004). Experimental and computational methods in cardiovascular fluid mechanics, *Ann. Rev. Fluid Mech.*, 36, 197–231.

Westerhoff, H. (2001). The silicon cell, not dead but live!, *Metabolic Eng.*, 3(3), 207–210.

Chapter 3

MATERIALS SIMULATION

Peter T. Cummings

3.1 Introduction

The discovery and adoption of new materials has, in extreme cases, been revolutionary, changed lives and the course of civilization, and led to whole new industries. Even our archeological classifications of prehistory (the stone, bronze, and iron ages) are named by the predominant materials of the day; the evolution of these materials (and their use in tools) is related to the predominant dwellings and societies of these ages (caves, farmsteads, and cities, respectively). By analogy, we might refer to recent history (the last half-century) as the plastics age followed by the silicon age. Today, we recognize a number of classes of materials, including but not limited to:

- smart/functional materials (whose physical and chemical properties are sensitive to, and dependent on, changes in environment, such as temperature, pressure, adsorbed gas molecules, stress, and pH);
- structural materials;
- electronic materials;
- nanostructured materials;
- biomaterials (materials that are biological in origin, such as bone, or are biomimetic and substitute for naturally occurring materials, such as materials for the replacement and/or repair of teeth).

Materials today have ubiquitous application in many industries (chemicals, construction, medicine, automotive, and aerospace, to name a few). Clearly, materials play a central role in our lives, and have done so since antiquity.

Given the importance of materials, understanding and predicting the properties of materials, either intrinsically or in structures, have long been goals of the scientific community. One of the key methodologies for this, materials simulation, encompasses a wide range of computational activities. Broadly defined, materials simulation is the computational solution to a set of equations that describe the physical and chemical processes of interest in a material. At the most fundamental level (often referred to as "full physics"), the equation to be solved is the Schrödinger equation (or its equivalent). If quantum degrees of freedom are unimportant, and atoms/molecules can be assumed to be in their ground state, then the forces between atoms/molecules can be described by simple functions (called forcefields), and the equations to be solved are Newton's equations or variants thereof (yielding molecular dynamics simulations). At longer time scales and larger spatial scales, mesoscale simulation methods such as dissipative particle dynamics (DPD) become the method of choice, while at scales near and at the macroscale, finite-element methods solving energy/mass/momentum partial differential equations are appropriate.

Full-physics materials simulations, when performed even on modest-scale structures (e.g., nanoparticles with several thousand atoms), have the ability to saturate with ease the largest available supercomputers (the so-called leadership-class computing platforms). Thus, full-physics (or first-principles) materials simulation of nanoscale and larger structures are among the applications driving the current progression toward petascale, to be followed by exascale, computing platforms.

In practice, materials simulators rarely have enough compute cycles available, even on petascale computers, to address at their desired level of accuracy the complex materials modeling problems that interest them. As a consequence, materials simulations almost always involve some tradeoff between accuracy (reflected in the methodology), size (number of atoms, or electrons, in the case of *ab initio* calculations), and duration. For example, first-principles methods, such as density functional theory (DFT), and molecular-orbital-based methods, such as Hartree–Fock (HF) and Møller–Plesset (MP) perturbation theory, range in their system-size dependence from N^3 to N^7, where N is the number of nuclei or outer-shell electrons. Computation times are clearly linear in the number of times steps simulated (or total simulated time); however, the number of times steps needed to bring a system to equilibrium can vary nonlinearly with the size of the system (e.g., structural relaxation times of linear polymers vary as the chain length squared). Thus, it is clear that materials simulations can rapidly escalate into computational grand challenges.

From among the available methods, the materials simulator must select the appropriate theory/method for given phenomenon and material; must be aware of the approximations made and how they effect outcome (error estimates); and ensure that the chosen methodology gives the right answer for the right reason (correct physics, correct phenomenon) (Carter, 2007). The wide range of methods available — even at a single level of description, such as the *ab initio* (also referred to as first-principles or full-physics) level — combined with limits on their applicability to different problems makes choosing the right theory a significant challenge. Additionally, for the novice user, or for the applied user interested primarily in pragmatically performing the most accurate and reliable calculation within available resources, there is little help available to navigate the minefield of available methods.

The situation is complicated even further by a lack of consensus or communal approach in the U.S. quantum chemistry and materials modeling communities. Unlike some other computational science communities, such as the climate modeling community, that have recognized the value of coming together to develop codes collaboratively and have coalesced around a limited number of community codes, the U.S. quantum chemistry and materials modeling communities remain divided, with many groups developing their own in-house codes. Such closed codes are inherently difficult to validate. There are exceptions. For example, in the biological simulation community, several codes — AMBER (see Amber Online), CHARMM (see Charmm Online), and NAMD (see Namd Online) — have emerged with significant funding from the National Institutes of Health (NIH) as community codes. The complexity of biological simulations has almost made this a necessity. The biological simulation community understands the strengths and weaknesses of the forcefields used in these codes, and the accuracy of the codes (for a given forcefield) has been subject to numerous consistency checks. The differences between these biological simulation codes often reduce to issues such as parallel performance and the availability of forcefield parameters for the system of interest.

This chapter addresses some of the issues and trends related to materials simulation identified during the WTEC panel's international assessment site visits. It begins with a brief discussion of the state of the art, then provides examples of materials simulation work viewed during the panel's international visits, and concludes with the findings of the panel related to materials simulation.

3.2 Current State of the Art in Materials Simulation

Materials simulation is an exceptionally broad field. To fully review the current state of the field is beyond the scope of this chapter; instead it provides references to recent reviews. For first principles methods, excellent overviews are provided by Carter (Carter, 2007; Huang and Carter, 2008); other aspects of hard materials modeling are reviewed by Chatterjee and Vlachos (2007) and Phillpot *et al.* (2007). Reviews of modeling methods for soft materials (i.e., polymers and polymer composites) are provided by Glotzer and Paul (2002), Theodorou (2007), Likos (2006), and Zeng *et al.* (2008).

Hattel (2008) reviews combining process-level simulation with materials simulation methods, where applicable, for engineering design. Integration of materials simulation methods into design, termed integrated computational materials engineering (ICME), is the subject of a 2008 report from the National Academy of Science and the National Academy of Engineering (NRC, 2008) that documents some of the achievements of ICME in industry and the prospects for the future of this field. Through documented examples, the report makes the case that investment in ICME has a 3:1 to 9:1 return in reduced design, manufacturing, and operating costs as a result of improved product development and faster process development. As just one example, the Virtual Aluminum Casting methodology (Allison *et al.*, 2006) developed at Ford Motor Company resulted in a 15–25% reduction in product development time, substantial reduction in product development costs, and optimized products, yielding a return of investment greater than 7:1.

Clearly, materials simulation methods, particularly when integrated with engineering design tools, can have significant economic impact. Equally significant is the role of materials simulation in national security; for example, materials simulation lies at the heart of the National Nuclear Security Administration's (NNSA's) mission to extend the lifetime of nuclear weapons in the U.S nuclear stockpile. Through ICME modeling on some of the world's largest supercomputers, the NNSA Advanced Simulation and Computing (ASC) program helps NNSA to meet nuclear weapons assessment and certification requirements through computational surrogates for nuclear testing and prediction of weapon performance.

3.3 Materials Simulation Code Development

Given the potential value of materials simulation, both for fundamental science (understanding the molecular origin of properties and dynamics) and for applications (as evidenced by the 2008 NRC report), and the ever-increasing complexity of systems that scientists and engineers wish to understand and model, development of materials simulation methods and tools is an activity engaged in worldwide. As noted in the introduction, within the United States (apart from notable exceptions described below) methods/tools development is largely individual-group-based. There are a number of factors that could account for this: (1) the tenure and promotion system in U.S. universities encourages individual achievement over collaboration during the first half-decade of a faculty member's career, thus establishing for many faculty a career-long focus on developing codes in competition with other groups; (2) funding agencies have tended not to fund code development activities *per se*, but rather to fund the solution of problems, a by-product of which may be a new method, tool, or code; (3) a corollary of this is that the person-power needed to perfect, distribute, and maintain tools/codes is generally not supported by funding agencies; and (4) within the United States, turning an in-house academic code into a commercial products is regarded as a successful outcome. Academic codes that have been commercialized are the original sources of almost all of the commercial materials and modeling codes available today. One might argue that the lack of federal agency funding for code development encourages an academic group to commercialize its code to create a revenue stream to fund distribution, support, and further development.

By contrast, in Japan and in Europe, there have been a number of community-based codes developed, many of which end up as open-source projects. In Japan, the multiscale polymer modeling code OCTA (http://octa.jp) was the result of a four-year multimillion-dollar government–industry collaboration to produce a set of extensible public-domain tools to model polymeric systems over many scales. In the UK during the 1970s, the predecessor of the Engineering and Physical Sciences Research Council (EPSRC) established Collaborative Computational Projects (CCPs, http://www.ccp.ac.uk/) with headquarters at Daresbury Laboratory to develop community-based codes for a wide range of simulation and physical sciences modeling applications (Allan, 1999). Among the codes developed by the CCPs are the atomistic simulation package DL_POLY (developed within CCP5), the electronic structure code GAMESS-UK (developed within CCP1), and the *ab initio* surface-properties-calculation program CASTEP (developed within CCP3). CASTEP is now marketed commercially worldwide by Accelrys (see Accelrys Online); thus, U.S. academic researchers must pay for using CASTEP, whereas in the UK, academic users have free access to CASTEP as a result of the role of CPP3 in its development. Other codes developed in European academic and/or government laboratory groups that are routinely used in the United States for materials modeling include VASP (see Vasp Online), CPMD (see Cpmd Consortium Online), SIESTA (see Siesta Online), and ESPRESSO (see Quantum Espresso Online). The development of community-based codes is evidently one area in which the United States trails other countries, particularly Japan and the EU.

Although fewer in number, there have been examples of codes developed within the United States, but in general, the circumstances under which they have been developed are unusual. For example, the development of the quantum chemistry program NWCHEM (Kendall *et al.*, 2000) was funded innovatively as part of the *construction* budget for the Environmental Molecular Science Laboratory (EMSL) at the Department of Energy's (DOE's) Pacific Northwest National Laboratory (PNNL). Maintaining the support for NWCHEM beyond the EMSL construction phase has proved challenging. The popular LAMMPS molecular dynamics (MD) code (Plimpton, 1995) was developed at Sandia National Laboratory (SNL) under a cooperative research and development agreement (CRADA) between DOE and Cray Research to develop codes that could take advantage of the Cray parallel computers of the mid-1990s; as the code is no longer supported by the CRADA or any other federal funding, its developer, Steve Plimpton, has placed it in the public domain under the GNU public license (http://www.gnu.org), and there is now a reasonably active open-source development project for LAMMPS. Part of the attraction of LAMMPS is its high efficiency on large parallel machines; additionally, the existence of converters from AMBER to/from LAMMPS and CHARMM to/from LAMMPS mean that biological simulations can be performed efficiently on large machines using LAMMPS as the MD engine but making use of the many analysis tools available for AMBER and CHARMM. These are exceptions rather than the rule, and in each case there are unique circumstances surrounding the development of the code.

The United States is headquarters to several of the commercial companies in materials modeling, such as Accelrys, Schrödinger, Spartan, and Gaussian; however, the viability of commercial ventures for chemicals and materials modeling has long been a subject of debate, largely driven by questions about whether the worldwide commercial market for chemicals and materials modeling is currently large enough to sustain multiple commercial ventures.

To understand how the different models for developing codes have evolved in the United States on the one hand, and Japan and Europe on the other, the WTEC panel made this subject one of the focuses of its site visits. One of the clear differences that emerged is funding. Japanese and European agencies actively fund large collaborative projects that have as their focus and outcome new simulation methodologies and codes. A secondary factor may be the tenure and promotion system in Japan and Europe, where faculty members are normally not subject to a long tenure tracks and often join the department and group in which they obtained their PhDs. While philosophically the WTEC panelists believe that the U.S. system, which makes such inward-focused hiring practices almost impossible, is superior in general (and is increasingly being adopted outside the United States), it is nevertheless evident that positive side effects of the Japanese/European system are continuity and maintaining of focus over the many years needed to develop a complex materials simulation capability.

3.4 Materials Simulation Highlights

The WTEC study panel visited a number of sites in Asia and Europe conducting materials simulations research, some of which are highlighted below. These highlights are not intended to be comprehensive — the panel visited more materials simulation sites than those listed below — nor is inclusion in this report meant to suggest that these sites conducted higher-quality research than other sites visited; rather, the intent is to choose a selection of sites that are representative of the range of activities observed during the WTEC study.

3.4.1 *Mitsubishi Chemical*

The Mitsubishi Chemical Group Science and Technology Research Center (MCRC), headquartered in Yokohama, Japan, is the corporate research center for Mitsubishi Chemical Corporation (MCC) one of three subsidiaries of Mitsubishi Chemical Holdings Company (MCHC). The majority of the research conducted by MCRC ($\sim$90%) is focused on healthcare and performance materials businesses.

MCRC staff use a variety of tools to design, construct, and operate optimal chemical and materials manufacturing processes. Approximately 90% of the tools used are commercial software packages (Aspen Tech and gPROMS for process-level modeling, STAR-CD and FLUENT for

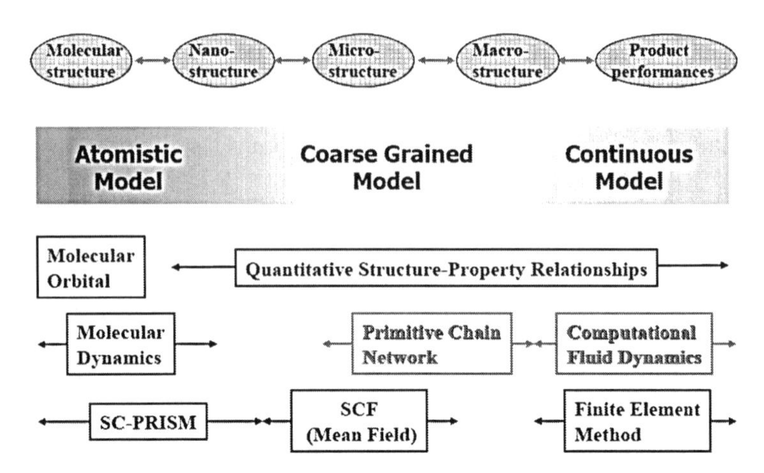

Fig. 3.1. Mitsubishi multiscale modeling approach for polymers.

computational fluid dynamics). The panel's hosts stated that MCC management has become comfortable with relying on predictions of simulation studies for design and optimization of chemical and materials manufacturing processes. In order to achieve its design goals, MCRC needs to couple these different codes together to create a multsicale model of an intended process. It achieves this goal by writing its own in-house codes to provide coupling between the commercial codes. MCRC supports an internal molecular modeling effort on polymers (illustrated in Fig. 3.1), which is linked to the process modeling efforts. For equilibrium properties (thermodynamics and structure), MCRC researchers have developed a reliable capability to relate molecular structure and properties, based on a combination of the molecular self-consistent polymer reference interaction site model (SC-PRISM) and self-consistent field (SCF) theories.

Of particular interest from a materials simulation point of view is the 17-member Computational Science Laboratory (CSL) headed by Shinichiro Nakamura. Using CSL-built in-house clusters of various sizes, the largest containing 800 CPUs, and the TSUBAME machine (http://www.gsic.titech.ac.jp), the CSL group uses *ab initio* and other methods for product design and solving process problems. In recent years, they have designed a robust (non-photodegrading) dye for use in printing (Kobayashi *et al.*, 2007), designed a high-quantum-yield yttrium oxysulfide phosphor for use in television sets (Mikami and Oshiyama, 1998, 1999, 2000; Mikami *et al.*, 2002), found an effective additive to improve the performance of Li-ion batteries (Wang *et al.*, 2001, 2002) and developed a new methodology for protein nuclear magnetic resonance (NMR) (Gao *et al.*, 2007) based on calculating the NMR shift using the fragment molecular orbital (FMO) methodology developed by Kazuo Kitaura for large-scale *ab initio* calculations. CSL researchers have contributed to community-based *ab initio* codes (Gonze *et al.*, 2002). The CSL performs fundamental, publishable research, as well as proprietary research that directly benefits MCC and MCHC.

3.4.2 *Toyota Central R&D Labs, Inc.*

Established in 1960 to carry out basic research for Toyota Group companies, the Toyota Central R&D Labs (TCRDL) performs a wide range of modeling activities. In the materials area, TCRDL is developing functional materials, including metallic composites, organic–inorganic molecular composites, and high-performance cell materials, as well as developing plastics and rubber recycling technologies. Funding for these activities is drawn from the Toyota Group corporations, and projects are selected from the technical needs of the Toyota Group and from proposals developed by the TCRDL staff. There are significant simulation efforts at TCRDL focused on the chemistry of combustion in internal combustion engines, on the multiscale modeling of fuel cells, on the performance of catalysts in oxidizing carbon monoxide, and on predicting the viscosity of synthetic lubricants. Most of the simulation work is performed on stand-alone workstations. However, if more computing power is needed, TCRDL has an NEC SX-5 on site and has mechanisms in place to buy computer time at the Earth Simulator. In terms of CPU time, TCRDL spends the most time in performing materials science simulations to uncover material structure–property relationships, followed in decreasing order by time spent on structural simulations of vibrations and acoustics, magneto-electric simulations in support of LIDAR (LIght Detection and Ranging) applications, and CFD/combustion simulations of internal combustion engines.

One particularly noteworthy project is TCRDL's effort to predict the viscosity of synthetic lubricants. Molecular dynamics simulations of the viscosity of known molecules with small simulation volumes could reproduce the trends observed in the laboratory, but the absolute value of the viscosity predicted by the initial simulation was incorrect by orders of magnitude because of the difference in the film thicknesses and shear rates of the experiment versus the simulation. Large-scale molecular dynamics simulations were then conducted on the Institute for Molecular Science

SR11000 in which the film thickness and shear rate in the simulation was comparable to the experimental measurements, and the two came into agreement.

TCRDL researchers propose to address in the future two computational materials grand challenges: to perform first-principles molecular dynamics simulations of fuel cells to understand the mechanisms of hydrogen atom transport within those systems, and to understand the nature of carbon–oxygen bonding and resonances in the vicinity of catalyzing surfaces.

3.4.3 *Joint Laboratory of Polymer Science and Materials, Institute of Chemistry, Chinese Academy of Sciences*

Founded in 2003 by Charles Han, formerly of the U.S. National Institute of Standards and Technology (NIST), the Joint Laboratory of Polymer Science and Materials (JLPSM) of the Institute of Chemistry, Chinese Academy of Sciences (ICCAS), has as its goal to unify the experimental, theoretical, and computational spectrum of polymer science and technology activities ranging from atomistic scale, to mesoscale materials characterization, to macroscale applied polymer rheology and processing. Between 1998 and 2004, the labs comprising the JLPSM published 1038 papers, were granted 121 patents, received four China science prizes, and presented 73 invited lectures at international conferences. For the JPPSM, 60–70% of student funding comes from the National Natural Science Foundation of China (the Chinese equivalent of the U.S. National Science Foundation), with almost none coming from the ICCAS. This is a different model than that used in the past. Increased funding is leading to 20–30% growth per year in students and faculty. Examples of materials simulations at the JLPSM include:

- molecular dynamics simulations of polyelectrolytes in a poor solvent (Liao *et al.*, 2003a,b)
- free-energy calculations on effects of confinement on the order-disorder transition in diblock copolymer melts
- organic electronic and photonic materials via *ab initio* computations using VASP, charge mobility for organic semiconductors (Wang *et al.*, 2007)
- organic and polymer light-emitting diodes (OLEDS and PLEDS) (Peng *et al.*, 2007)

- soft matter at multiple length scales, including molecular dynamics and coarse-grained models
- molecular dynamics-Lattice Boltzmann hybrid algorithm, microscopic, and mesoscale simulations of polymers
- theory of polymer crystallization and self-assembly

Several of the JLPSM researchers develop their own codes and in some cases have distributed them to other research groups.

One notable large activity is overseen by Charles Han. The project focuses on integrated theory/simulation/experiment for multiscale studies of condensed phase polymer materials. It is funded with 8 million RMB[1] over four years, and has 20 investigators, with roughly 100,000 RMB per investigator. This is a unique and challenging program, inspired by the Doi project (http://octa.jp) in Japan, but it goes substantially beyond that project by closely integrating experiment, both as a validation tool and to supply needed input and data for models.

3.4.4 *Materials Simulation Code Development in the UK*

The UK has a relatively long tradition of developing materials simulation codes. Of particular note are the Collaborative Computational Projects (CCPs, http://www.ccp.ac.uk) hosted at Daresbury Laboratory (see Daresbury Laboratory site report for a full listing of the CCPs). The CCPs consist of Daresbury staff along with university collaborators whose joint goal is the development and dissemination of codes and training for their use. The majority, but not all, of the CCPs are involved in materials simulation. "CCP5" — the CCP for Computer Simulation of Condensed Phases — has produced the code DL_POLY, a popular molecular dynamics simulation package. Other codes produced or maintained by the CCPs are GAMESS-UK (a general purpose *ab initio* molecular electronic structure program, http://www.cfs.dl.ac.uk/), CASTEP (a density-functional-theory-based code to calculate total energies, forces optimum geometries, band structures, optical spectra, and phonon spectra, http://www.castep.org), and CRYSTAL (a Hartree–Fock and density-functional-theory-based code for computing energies, spectra and optimized geometries of solids, http://www.crystal.unito.it). CASTEP can also be

[1]At the time of writing, $US 1 (USD) is approximately 7 Chinese yuan (RMB). RMB is an abbreviation for renminbi, the name given to the Chinese currency, for which the principal unit is the yuan.

used for molecular dynamics with forces calculated from density functional theory (DFT). Recent changes in the funding landscape in the UK has resulted in the CCPs being responsible primarily for code maintenance and distribution; they no longer have the resources to support significant new code development.

CASTEP (Cambridge Sequential Total Energy Package) was primarily developed by Mike Payne of Cambridge University. With coworkers Peter Haynes, Chris-Kriton Skylaris, and Arash A. Mostofi, Payne is currently developing a new code, ONETEP (order-N electronic total energy package, http://www.onetep.soton.ac.uk) (Haynes *et al.*, 2006). ONETEP achieves order-N scaling (i.e., computational cost that scales linearly, $O(N)$, in the number of atoms), whereas DFT methods typically scale as the cube of the system size, $O(N^3)$, and molecular orbital methods as $O(N^5)$–$O(N^7)$. Examples are provided by Skylaris *et al.* (2008), and include the application to DNA shown in Fig. 3.2.

As valuable as the CCPs have been, there are concerns for the future. The Cambridge researchers expressed a concern echoed by others, namely that of the declining programming skills of incoming graduate students.

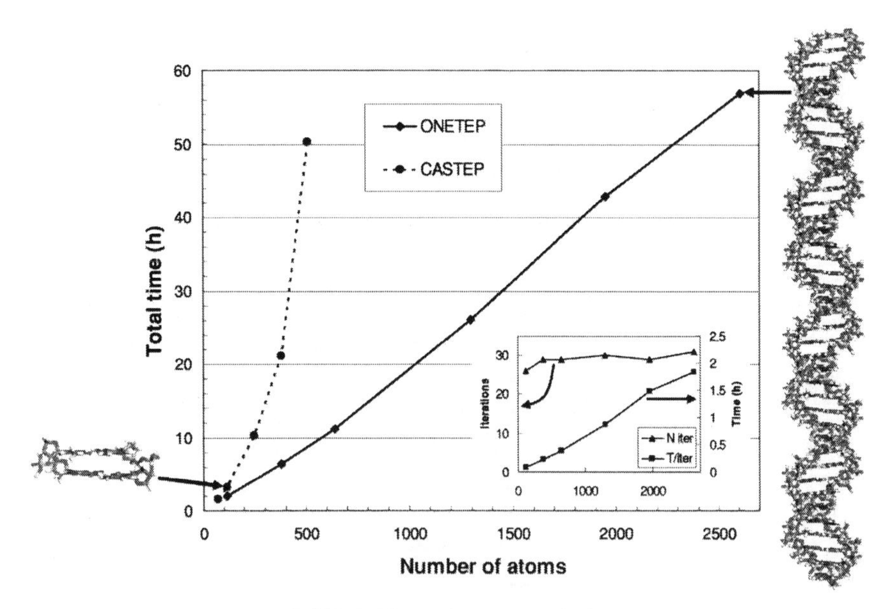

Fig. 3.2. Linear scaling of ONETEP total energy calculation on DNA compared to CASTEP. Total time for the calculation is plotted against number of atoms. Calculations are performed on a 64-processor computer (from Skylaris *et al.*, 2008).

This is being addressed to some degree by the educational programs at centers for scientific computing established at the universities of Warwick (http://www2.warwick.ac.uk) and Edinburgh (http://www.epcc.ed.ac.uk), and workshops hosted by the CCPs at Daresbury Laboratory. Additionally, there was considerable anecdotal evidence that the UK's Engineering and Physical Sciences Research Council (http://www.epsrc.ac.uk) is funding the CCPs primarily for code maintenance and training, not for code development. A long-term problem for the development of materials modeling codes is that the expectation of funding agencies and university administrators for "sound-bite science," publishable in the highest-profile journals. Software engineering is incremental by nature, and these "sound-bite science" expectations mitigate against materials software engineering receiving its due credit. Finally, it is too early to tell whether the reorganization of UK's science laboratories under the new Science and Technology Facilities Council (http://www.scitech.ac.uk/) will be beneficial for the development of materials modeling codes.

3.4.5 *Fraunhofer Institute for the Mechanics of Materials*

The Fraunhofer Institute for Mechanics of Materials (Institut Werkstoffmechanik, IWM), with locations in Freiburg and Halle/Saale, Germany, is one of the Fraunhofer Institutes that constitute the largest applied research organization in Europe, with a research budget of €1.3 billion and 12,000 employees in 56 institutes. The institutes perform their research through "alliances" in Microelectronics, Production, Information and Communication Technology, Materials and Components, Life Sciences, Surface Technology and Photonics, and Defense and Security. Financing of contract research is by three main mechanisms: institutional funding, public project financing (federal, German Länder, EU, and some others), and contract financing (industry). The IWM has 148 employees (with another 75 based at its other campus in Halle), and a €15.5 million budget, with 44% of the budget coming from industry and 25–30% from the government. The IWM has seen significant growth in recent years (10% per year), a figure that is currently constrained by the buildings and available personnel. At the IWM, 50% of the funding supports modeling and simulation (a number that has grown from 30% five years ago). The IWM has 7 business units: (1) High Performance Materials and Tribological Systems, (2) Safety and Assessment of Components, (3) Components in Microelectronics, Microsystems and Photovoltaics, (4) Materials-Based

Process and Components Simulation, (5) Components with Functional Surfaces, (6) Polymer Applications, and (7) Microstructure-Based Behavior of Components.

The IWM has a world-class effort in applied materials modeling. For example, it has developed microscopic, physics-based models of materials performance and then inserted these subroutines into FEM codes such as LSDYNA, ABAQUS, and PEMCRASH to develop multi-physics-based automobile crash simulations. The IWM enjoys strong collaborations with German automobile companies, which support this research. Other research areas include modeling of materials (micromechanical models for deformation and failure, damage analysis), simulation of manufacturing processes (pressing, sintering, forging, rolling, reshaping, welding, cutting), and simulation of components (prediction of behavior, upper limits, lifetime, virtual testing).

The IWM employs mostly externally developed codes for its simulations (with some customized in-house codes that couple to the externally developed codes). All the codes run on parallel machines. At the largest (non-national) scale, the Fraunhofer institutes have a shared 2000-node cluster, of which capacity the IWM uses approximately 25%.

IWM staff find it challenging to find materials scientists with a theoretical background strong enough to become effective simulators. The typical training is in physics, and materials knowledge must then be acquired through on-site training. Currently it is difficult to obtain staff skilled in microstructure-level simulation, as they are in high demand by industry.

3.4.6 *Energy Applications of Materials at Daresbury Laboratory*

The researchers located in the materials CCPs at Daresbury Laboratory have an overarching materials goal of developing new materials, and modeling existing materials, for energy applications. Examples of energy-relevant materials under study at Daresbury Laboratory include fuel cells, hydrogen storage materials, nuclear containment systems, photovoltaics, novel electronics and spintronics systems (including quantum dots (Tomić, 2006), and solid state battery electrodes. Specific properties of interest include structure and composition, thermodynamics and phase stability, reaction kinetics and dynamics, electronic structure (correlation), and dynamics. The toolkit for studying these properties at Daresbury Laboratory consists of the CCP codes DL_POLY, CASTEP, CRYSTAL,

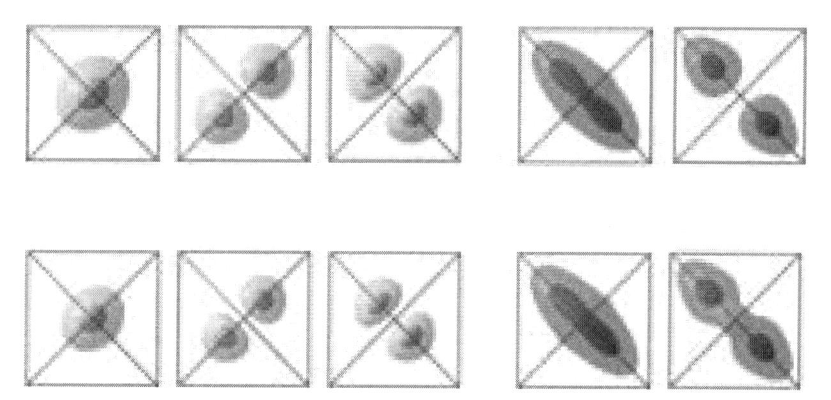

Fig. 3.3. Lowest 3-electron wave functions and highest 2-hole wave functions of InAs (top row) and InGaAsN (bottom row) quantum dots grown on GaAs substrate calculated using Daresbury kkpw code (Tomić, 2006).

and a parallel implementation of the kppw continuum electronic structure code (see Fig. 3.3). Using combinations of these codes, the Daresbury Laboratory researchers have studied radiation damage in nuclear reactors, performed atomistic simulations of resistance to amorphization by radiation damage, and simulated radiation damage in gadolinuim pyrochlores.

3.5 Summary of Key Findings

SBE&S has long played a critical role in materials simulation. As the experimental methodologies to synthesize new materials and to characterize them become increasingly sophisticated, the demand for theoretical understanding is driving an ever greater need for materials simulation. This is particularly so in nanoscience, where it is now possible to synthesize an infinite variety of nanostructured materials and combine them into new devices or systems with complex nanoscale interfaces. The experimental tools of nanoscience — such as scanning probes, neutron scattering, and various electron microscopies — all require modeling to understand what is being measured. Hence, the demand for high-fidelity materials simulation is escalating rapidly, and the tools for verification of such simulations are becoming increasingly available.

The WTEC panel's key findings are summarized below:

- Computational materials science and engineering is changing how new materials are discovered, developed, and applied. We are in an extra ordinary period in which the convergence of simulation and experiment at the nanoscale is creating new opportunities for materials simulation, both in terms of new targets for study and in terms of opportunities for validation.

- World-class research in all areas of materials simulation areas is to be found in the United States, Europe, and Asia.
  - Algorithm innovation takes place primarily in the United States and Europe, some in Japan, with little activity to date in this area in China.
  - China has been mostly concerned with applications in the past; however, as part of the Chinese government current emphasis on supporting creativity in the both the arts and sciences, one can anticipate increased activity in this area in China.
- There is a rapid ramping-up of materials simulation activities in some countries, particularly China and Germany.
- There is an extraordinary revival of science in Germany, tied to recent increases in the Deutsche Forschungsgemeinschaft (DFG, German Research Foundation, the equivalent of the U.S. National Science Foundation) budget. In the most recent years, DFG funding jumped over 20%, following on three years of increases in the 5–10% range.
- There is much greater collaboration among groups in code development in Europe compared to the United States. There appear to be several reasons for this:
  - The U.S. tenure process and academic rewards systems mitigate against collaboration.
  - Funding, promotion, and awards favor high-impact science (publications in *Nature* and *Science*, for example), while the development of simulation tools is not considered to be high-impact science. There are numerous counter-examples, in which simulation software has been critical in scientific breakthroughs, but typically the breakthrough science comes a long time after a software tool has been developed, adopted widely, and validated. In other words, the payback on the development of a new simulation tool can be a decade or more after it is first developed.
- Training is a critical issue worldwide, as undergraduate students in the physical and chemical sciences and engineering are increasingly illiterate in computer programming. The situation is perhaps most acute of all in the United States.

- Funding of long-term (5 years at the very minimum), multidisciplinary teams is crucial to enable the United States to develop key materials simulation codes that will run on the largest U.S. computing platforms and on future many-core processors. Such codes will be needed to explore the next generation of materials and to interpret experiments on the next generation of scientific instruments.
- The United States is at an increasingly strategic disadvantage with respect to crucial, foundational codes, as it has become increasingly reliant on codes developed by foreign research groups. Political considerations can make those codes effectively unavailable to U.S. researchers, particularly those in Department of Defense laboratories.
- The utility of materials simulation codes for practical application would be enhanced dramatically by the development of standards for interoperability of codes. This would be similar to the CAPE-OPEN effort undertaken in the chemical process simulation (computer-aided process engineering, or CAPE) field to develop interoperability standards in that field (http://www.colan.org).

References

Accelrys Online (2010). http://accelrys.com/products/materials-studio/modules/CASTEP.html.

Allan, R. J., ed. (1999). *High Performance Computing* (Kluwer Academic/Plenum Press, New York).

Allison, J., Li, M., Wolverton, C. and Su, X. M. (2006). Virtual aluminum castings: an industrial application of ICME, *JOM*, 58(11), 28–35.

Amber Online (2010). http://ambermd.org.

Carter, E. A. (2007). Status and challenges in quantum mechanics based simulations of materials behavior, *Review of Simulation-Based Engineering and Science Research in the United States, 1–2 November, Arlington, VA* (WTEC, USA).

Charmm Online. http://www.charmm.org.

Chatterjee, A. and Vlachos, D. G. (2007). An overview of spatial microscopic and accelerated kinetic Monte Carlo methods, *J. Computer-Aided Mater. Des.*, 14(2), 253–308.

Cpmd Consortium Online, http://www.cpmd.org.

Gao, Q., Yokojima, S., Kohno, T., Ishida, T., Fedorov, D. G., Kitaura, K., Fujihira, M. and Nakamura, S. (2007). Ab initio NMR chemical shift calculations on proteins using fragment molecular orbitals with electrostatic environment, *Chem. Phys. Lett.*, 445(4–6), 331–339.

Glotzer, S. C. and Paul, W. (2002). Molecular and mesoscale simulation of polymers, *Ann. Rev. Mater. Res.*, 32, 401–436.

Gonze, X., Beuken, J. M., Caracas, R., Detraux, F., Fuchs, M., Rignanese, G.-M., Sindic, L., Verstraete, M., Zerah, G., Jollet, F., Torrent, M., Roy, A., Mikami, M., Ghosez, P., Raty, J.-Y. and Allan, D. C. (2002). First-principles computation of material properties: the ABINIT software project, *Comput. Mater. Sci.*, 25(3), 478–492.

Hattel, J. H. (2008). Integrated modelling in materials and process technology, *Mater. Sci, Technol.*, 24(2), 137–148.

Haynes, P. D., Skylaris, C.-K., Mostofi, A. A. and Payne, M. C. (2006). ONETEP: linear-scaling density-functional theory with local orbitals and plane waves, *Phys. Stat. Solid. B*, 243, 2489–2499.

Huang, P. and Carter, E. A. (2008). Advances in correlated electronic structure methods for solids, surfaces, and nanostructures, *Ann. Rev. Phys. Chem.*, 59, 261–290.

Kendall, R. A., Apra, E., Bernholdt, D. E., Bylaska, E. J., Dupuis, M., Fann, G. I., Harrison, R. J., Ju, J., Nichols, J. A., Nieplocha, J., Straatsma, T. P., Windus, T. L. and Wong, A. T. (2000). High performance computational chemistry: an overview of NWChem a distributed parallel application, *Comput. Phys. Comm.*, 128(1–2), 260–283.

Kobayashi, T., Shiga, M., Murakami, A. and Nakamura, S. (2007). Ab initio study of ultrafast photochemical reaction dynamics of phenol blue, *J. Am. Chem. Soc.*, 129(20), 6405–6424.

Liao, Q., Dobrynin, A. V. and Rubinstein, M. (2003a). Molecular dynamics simulations of polyelectrolyte solutions: nonuniform stretching of chains and scaling behavior, *Macromolecules*, 36(9), 3386–3398.

Liao, Q., Dobrynin, A. V. and Rubinstein, M. (2003b). Molecular dynamics simulations of polyelectrolyte solutions: osmotic coefficient and counterion condensation, *Macromolecules*, 36(9), 3399–3410.

Likos, C. N. (2006). Soft matter with soft particles, *Soft Matter*, 2(6), 478–498.

Mikami, M., Nakamura, S., Itoh, M., Nakajima, K. and Shishedo, T. (2002). Lattice dynamics and dielectric properties of yttrium oxysulfide, *Phys. Rev. B*, 65(9), 094302.1–094302.4.

Mikami, M. and Oshiyama, A. (1998). First-principles band-structure calculation of yttrium oxysulfide, *Phys. Rev. B*, 57(15), 8939–8944.

Mikami, M. and Oshiyama, A. (1999). First-principles study of intrinsic defects in yttrium oxysulfide, *Phys. Rev. B*, 60(3), 1707–1715.

Mikami, M. and Oshiyama, A. (2000). First-principles study of yttrium oxysulfide: bulk and its defects, *J. Lumin.*, 87(9), 1206–1209.

Namd Online, http://www.ks.uiuc.edu/Research/namd.

National Research Council, Committee on Integrated Computational Materials Engineering (2008). *Integrated Computational Materials Engineering: a Transformational Discipline for Improved Competitiveness and National Security* (The National Academies Press, Washington, DC).

Peng, Q., Yi, Y. and Shuai, Z. G. (2007). Excited state radiationless decay process with Duschinsky rotation effect: formalism and implementation, *J. Chem. Phys.*, 126(11), 114302–114308.

Phillpot, S. R., Sinnott, S. B. and Asthagiri, A. (2007). Atomic-level simulation of ferroelectricity in oxides: current status and opportunities, *Ann. Rev. Matl. Rsch.*, 37, 239–270.

Plimpton, S. (1995). Fast parallel algorithms for short-range molecular dynamics, *J. Comput. Phys.*, 117(1), 1–19.

Quantum Espresso Online, http://www.quantum-espresso.org.

Siesta Online, http://www.icmab.es/siesta.

Skylaris, C.-K., Haynes, P. D., Mostofi, A. A. and Payne, M. C. (2008). Recent progress in linear-scaling density functional calculations with plane waves and pseudopotentials: the ONETEP code, *J. Phys.-Condensed Matt.*, 20, 064209.

Theodorou, D. N. (2007). Hierarchical modelling of polymeric materials, *Chem. Eng. Sci.*, 62(21), 5697–5714.

Tomić, S. (2006). Electronic structure of InGaAsN/GaAs(N) quantum dots by 10-band kp theory, *Phys. Rev. B*, 73, 125348.

Vasp Website, http://cms.mpi.univie.ac.at/vasp.

Wang, L. J., Peng, Q., Li, Q. K. and Shuai, Z. G. (2007). Roles of inter- and intramolecular vibrations and band-hopping crossover in the charge transport in naphthalene crystal, *J. Chem. Phys.*, 127(4), 044506–044509.

Wang, Y. X., Nakamura, S., Tasaki, K. and Balbuena, P. B. (2002). Theoretical studies to understand surface chemistry on carbon anodes for lithium-ion batteries: how does vinylene carbonate play its role as an electrolyte additive? *J. Am. Chem. Soc.*, 124(16), 4408–4421.

Wang, Y. X., Nakamura, S., Ue, M. and Balbuena, P. B. (2001). Theoretical studies to understand surface chemistry on carbon anodes for lithium-ion batteries: reduction mechanisms of ethylene carbonate, *J. Am. Chem. Soc.*, 123(47), 11708–11718.

Zeng, Q. H., Yu, A. B. and Lu, G. Q. (2008). Multiscale modeling and simulation of polymer nanocomposites, *Prog. Polym. Sci.*, 33(2), 191–269.

Chapter 4

ENERGY AND SUSTAINABILITY

Masanobu Shinozuka

4.1 Introduction

The major energy problem the world faces at this time is the finite availability of fossil fuel. Therefore, energy conservation and the development of reliable clean energy sources should be the main focus of the future research effort. This observation is prompted by statistics prepared by the U.S. Department of Energy which demonstrate that combined energy consumption related to buildings and transportation systems amounts to an overwhelming 75% of the total U.S. consumption. These systems represent backbones of infrastructure systems busily supporting the modern built environment, and yet they can be made substantially more energy-efficient by making use of new materials and by means of new methods of construction, retrofit, operation, and maintenance.

Less use of fossil fuel diminishes emission of CO_2 and thus lowers its impact on global warming, assuming that acute correlation exists between them as proposed by some researchers. What is even more important, however, is the fact that the less we consume fossil fuel now, the more time we gain for the development of new, reliable, and substantial sources of clean energy before the arrival of Doom's Day when we deplete all economically affordable fossil fuel on the Earth.

Nuclear power is such a "clean energy" source, although its operational risk and reliability and the treatment of nuclear waste remain the subject of controversy. Biomass energy also appears to serve as a robust clean energy source, as demonstrated by Brazil producing large amounts of ethanol from sugar cane. Other sources such as solar, wind, and geothermal should also be pursued at a more accelerated pace. The development of renewable energy is too slow and provides too small percentage of the energy supply, although some future estimates are more optimistic. Also, we certainly hope

Table 4.1. Alternative energy sources & efficiencies: actions to cut CO_2 emissions by 1 Gt/year.[a]

Today's technology	Actions that provide 1 Gigaton/year of mitigation
Coal-fired power plants	1000 "zero-emission" 500-MW coal-fired power plants[b]
Nuclear	500 new nuclear power plants, each 1 GW in size[b]
Car efficiency	1 billion new cars at 40 miles per gallon (mpg) instead of 20 mpg
Wind energy	50 times the current global wind generation[b]
Solar photovoltaics (PV)	1000 times the current global solar PV generation[b]
Biomass fuels from plantations	15 times the size of Iowa's farmland
CO_2 storage in new forests	30 times the size of Iowa's farmland to new forest

[a] *Source*: http://genomics.energy.gov. Emission of CO_2 in 2010 is estimated to be 27.7 Gigaton (Almanacs).
[b] In lieu of coal-fired plants without CO_2 capture and storage.

that an economically practical hydrogen fuel cell will be delivered in more opportune time.

Table 4.1 shows an interesting comparison of what it takes to achieve annual reduction of one gigaton of CO_2 gas emission (about 3% of total global annual emission) by utilizing alternative energy sources or by means of some emission control technologies. Clearly, some of these technologies entail various socio-economic and/or geo-political repercussions, and others potentially interfere with the balance of the very natural environment that is to be protected. This shows that optimal selection of alternative energy sources requires complex socio-economic balancing act.

In this chapter, the subject matter of "Energy and sustainability" is considered from two points of view. In the first, the issue of the energy and its sustainability is considered as a set, and hence the development and utilization of alternative energy sources play an important role in this consideration. The second focuses on the issue of sustainability of the built environment, particularly infrastructure systems that must remain resilient under natural and man-made hazards such as earthquakes and malicious attacks, respectively. This observation is summarized as follows.

- *Energy and sustainability*: timely development of renewable and alternative energy is critical to America's energy security and sustainable adaptation to global climate change (U.S. Department of Energy).
- *Infrastructure development and sustainability*: "sustainable development is the challenge of meeting human needs for natural resources ...

while conserving and protecting environmental quality and the natural resource base essential for future development" (ASCE Policy 418, as first defined in November 1996; http://pubs.asce.org/magazines/ascenews/2008/Issue_01-08/article3.htm).

- *System-specific sustainability (a necessary condition)*: a system is sustainable under natural, accidental, and man-made hazards if it is designed sufficiently resilient relative to the return period of the extreme events arising from these hazards (Shinozuka, 2009).

The purpose of this chapter is to evaluate the state of the research in the world on energy and sustainability in the context of these definitions from the above-mentioned three points of view, specifically focusing on the research that utilizes simulation-based engineering and science technologies (SBE&S).

4.2 SBE&S Research Activities in North America

4.2.1 *Analysis of Energy and CO_2 Emission*

CO_2 emission has been one of the major concerns of the electric power industry; the Electric Power Research Institute (EPRI) carried out a detailed "Prism" analysis that projected by simulation (in a broad sense) the reduction of CO_2 emission that is achievable beyond 2020 and into 2030. The reduction contributions arising from different technologies are color-coded. These technologies include efficiency in generation, use of renewables, nuclear generation, advanced coal generation, carbon capture and storage, plug-in hybrid electronic vehicles, and distributed energy sources; this shows that carbon capture and storage provide the most significant reduction. The "Prism" analysis appears to include evaluation of the effectiveness of suppressing CO_2 emission for each technology. It would be highly useful if the details of this analysis can be made available to the research community.

4.2.2 *Modeling System Reliability of Electric Power Networks*

Analytical modeling and assessment of electric power system reliability have their origins in North America, as represented by a number of books authored by Endrenyi (1978), Billinton and Allan (1988), Billinton and Li (1994), and Brown (2002). Particularly, the last two books describe how to use Monte Carlo Methods for reliability assessment. In a recent paper by Zerriffi *et al.* (2007), the authors indicate that they augmented the traditional Monte Carlo reliability modeling framework and compared the performance of centralized to distributed generation (DG) systems under various levels of conflict-induced stresses. They concluded that DG systems are significantly more reliable than centralized systems, and when whole-economy costs are considered, they are also more economical. These simulations are carried out on less complex power network models or IEEE reliability test systems.

An analytical model was developed, improved, and expanded over the last decade by Shinozuka *et al.* for prediction of the seismic performance of power transmission networks under probabilistically defined seismic hazards. The most recent version is summarized in a paper by Shinozuka *et al.* (2007a). The model is sophisticated and heavily probabilistic and complex, requiring extensive use of Monte Carlo simulation for the prediction of system performance. The major advantage of using Monte Carlo simulation is that the system can be upgraded and retrofitted "virtually" and then subjected to various probabilistic hazard scenarios. This allowed the researchers to identify the most effective method of retrofit under probabilistic hazard scenarios in terms of shifting of risk curves. The simulation method is applied for transmission network systems using the networks of the Los Angeles Department of Water and Power (LADWP) and of Memphis Light, Gas, and Water (MLGW) as testbeds. The seismic performance in these examples is measured in terms of the remaining power supply capability as a percentage of the intact system capability. The seismic hazard is given by an ensemble of scenario earthquakes consistent with hazard definition of the U.S. Geological Survey. For this reason, the performance is given by a risk curve for LADWP's power system, in which only transmission-sensitive equipment (transformers, circuit breakers, disconnect switches, and buses) at receiving stations are considered for the failure analysis under seismic conditions. The risk curve indicates the annual exceedance probability that the network will suffer from more than a certain level of loss in power supply capability as a function of that level. The same model also demonstrated its ability to reproduce the actual recovery process of the damaged transmission network. Indeed, a good agreement was observed between actual and simulated states of recovery of the LADWP power transmission network after the 1994 Northridge earthquake (Shinozuka *et al.*, 2007a) showing the model's ultimate robustness.

4.2.3 *Modeling Civil Infrastructure Systems Resilience and Sustainability*

For the purpose of improving resilience and sustainability of civil infrastructure systems under natural and man-made hazards, retrofitting of the systems is often carried out at the expense of significant retrofit cost. Seismic retrofitting is a typical example. In the lack of well-established quantitative definitions for resilience and sustainability at present, a reasonable question that can be raised is if the retrofit is cost-effective, instead of asking if the structural systems increased their resilience and/or sustainability. In recent years, the California Department of Transportation (Caltrans') engaged in and completed seismic retrofit of its bridges by means of steel-jacketing the bridge columns. The same question of cost-effectiveness was raised here, and indeed, a benefit–cost analysis was performed (Zhou *et al.*, 2007). The tasks required in this analysis consist of the sequential modules, which are in essence the same as the task modules used for the development of risk curves associated with the power transmission network. The difference primarily arises from the fact that in this case the risk curve is developed as a function of the social cost associated with drivers' delay plus daily loss of opportunity cost, all measured in hour units. This definition of social cost places an emphasis on the Caltrans' responsibility as the major stakeholder for the degradation of the highway network due to seismically induced bridge damage. The social cost thus given in terms of hours of delay and opportunity loss is converted to equivalent monetary value for the benefit–cost analysis (average hourly wages of $21.77 for Los Angeles is used, per 2005 U.S. Department of Labor data). In this work, the authors introduced a network-wide probabilistic bridge repair model and simulated the resulting recovery of the network capability. This made it possible to trace the daily social cost, as a function of the elapsed time, taking into account the corresponding progress in the recovery effort. The total expected annual social cost of the network under the given seismic hazard could then be estimated.

The difference between the total annual social cost when no bridges are retrofitted and when all the bridges are retrofitted represents the expected annual cost avoided due to the retrofit — or benefit resulting from the retrofit. Similarly, expected annual bridge restoration cost avoided can be computed. The total expected annual cost avoided (sum of the social cost avoided and bridge restoration cost avoided) over the expected life (assumed to be 50 years) of the network is converted under a certain discount rate to the present value of the benefit of the retrofit. This value divided by

Table 4.2. Cost-effectiveness evaluation summary.

Discount rate	Benefit/cost ratio	Cost-effectiveness[a]
3%	4.39	No
5%	3.23	Yes
7%	2.36	Moderate

[a]R is the benefit/cost ratio, and the level of cost effectiveness is characterized as No, $R < 1.5$; Moderate, $1.5 \leq R < 2.5$; Yes, $R \geq 2.5$.

the retrofit cost indicates the benefit–cost ratio. This ratio is computed, in this example, under the discount rate of 3%, 5%, and 7% for the network life horizon of 50 years, as shown in Table 4.2, which shows that the cost-effectiveness of total bridge retrofitting can be rated as no, yes, and moderate under the discount rates of 3%, 5% and 7% respectively.

The meteorological conditions under global warming could increase the occurrence frequency and intensity of each adverse event and, therefore, the increased risk to the built environment. Developing new retrofit strategies and design guidelines should reduce or at least contain the damage risk of the built environment at an acceptable level. Typically, retrofitting of existing levy systems to reduce the damage risk is only one of many engineering problems in this case. An example of infrastructure failure due to a weather-related extreme event is studied by Sanders and Salcedo (2006). In this case, a scenario is considered such that an aging earth dam is breeched and fails under heavy rain. This floods the down-stream areas and inundates residential areas. Thus, the heavy rain causes a multiple hazard: (1) possible human casualty, (2) loss of property, (3) disruption of surface transportation by inundation, and (4) blackout of utility networks due to the malfunction of their control equipment being installed in the lower areas or in basements within utility facilities.

4.3 SBE&S Research Activities in Asia

4.3.1 *Modeling System Reliability of the Electric Power Network*

4.3.1.1 *National Center for Research on Earthquake Engineering (NCREE), Taipei Taiwan*

The Taiwan Power Company (Taipower) system suffered from devastating damage from the 1999 Chi-Chi earthquake (Mw = 7.2). The damage was widespread throughout the system, including high-voltage transmission lines and towers, power generating stations, substations, and foundation structures. Drs. Chin-Hsiung Loh and Karl Liu of NCREE in collaboration with Taiwan Power Co. were instrumental in developing a computational system not only for modeling high-voltage transmission systems including transmission lines and towers, but also for modeling the entire system, including power-generating stations. The LADWP study reviewed earlier, considered primarily receiving station equipment (transformers, disconnect switches, and circuit breakers) to be seismically vulnerable on the basis of past earthquake experience in California. In this context, NCREE's computer code is able to simulate a broader variety of seismic scenarios. For example, their code can demonstrate by simulation that higher level of seismic reliability can be achieved for a network with a distributed rather than centralized power generation configuration under the same maximum power output.

4.3.1.2 *Central Research Institute for Electric Power Industry (CRIEPI), Tokyo, Japan, and Chugoku Electric Power, Hiroshima, Japan*

Dr. Yoshiharu Shumuta of Japan's CRIEPI (Central Research Institute for Electric Power Industry) has developed a support system for Chugoku Electric Power to be able to implement efficient disaster response and rapid system restoration during and after the extreme wind conditions of Japan's ubiquitous typhoons (see Asia Site Reports on the International Assessment of Research and Development in Simulation-Based Engineering and Science website, http://www.wtec.org/private/sbes/. The required password is available from WTEC upon request). While, as in the case of FEMA HAZUS (the U.S. Federal Emergency Management Agency's software program for estimating potential losses from disasters), the wind-induced damage can be correlated with wind speed, direction, and duration. The CRIEPI's tool is more advanced in that it uses early prediction by the weather bureau for these wind characteristics. This will provide critically needed lead time for emergency response and significantly enhance the effectiveness of response and restoration. Dr. Shumuta was also successful in instrumenting electric poles with accelerometers. This allows assessing the integrity of network of electric wires serving as part of the distribution system (as opposed to transmission system). This is the first time the network of electric wires is instrumented for the purpose of system integrity under earthquakes. Dr. Shumuta contends that the same instrumentation is useful for assessment of integrity of the system under Typhoons. Simulation models of distribution network integrity, when developed, can be verified with the observations obtained from this type of measurements.

4.3.2 *Modeling Civil Infrastructure Systems Resilience and Sustainability*

4.3.2.1 *Pusan Port Authority, Pusan, Korea*

The 1995 Hyogoken-Nanbu (Kobe) earthquake ($M = 6.8$) caused serious damage to Kobe, Japan, port facilities, proving that the facilities were not quite resilient to an earthquake of this magnitude, and even worse, that they do not appear to be very sustainable more than one decade after the event. This observation prompted Dr. Ung Jin Na at Pusan Port Authority (Korea) to develop a computer simulation tool that will enable the authority to carry out a probabilistic assessment of seismic risk (and the risk from other natural hazards such as typhoons) that the facilities are likely to be subjected to. Major ports, particularly seaports, are important nodes of transportation networks providing shipping and distribution of cargos via water. Handling 97% of international cargo through seaports, port's down time results in severe regional and national economic losses. Unfortunately, port facilities are very vulnerable to earthquakes, being often located near to seismic faults and built on liquefaction-sensitive fill or soft natural materials. The most prominent highlight in Dr. Na's simulation work is the use of the spectral representation method (Shinozuka and Deodatis, 1996) of random multidimensional simulation for liquefaction analysis. The method is applied to the finite difference liquefaction analysis of saturated backfill soil whose key mechanical properties are modeled as a two-dimensional random field. A rather dramatic effect of the random field assumption is quite clearly observed when compared with the more subdued liquefaction behavior under the assumption of uniform material properties. The excess pore pressure distributions in the different 2D samples of this random field are obviously quite different from each other. Only with this irregular distribution of pressure, as simulated, was it possible to explain actual observation that some caissons displaced under the earthquake an unusually large distance of 5.5 m displacement, as opposed to the 2.5 m experienced by most of them, and that two adjacent caissons have significantly different values of outward displacements (Na *et al.*, 2008). For example a critical infrastructure system, in this case the

Kobe port system, demonstrated a reasonable resilience by restoring the cargo handling capability within about a year and half and yet its world ranking in cargo handling degraded from the fifth place before the event to the 35th place in 2004 because of the shifting of the cargo business to other ports while the Kobe port is struggling with the restoration. This represents an example which shows that an apparently reasonable resilience capability does not guarantee the system sustainability (see the description of System-specific sustainability in the introduction of this chapter).

4.3.2.2 *Disaster Control Research Center, School of Engineering, Tohoku University, Japan*

Professor Shunichi Koshimura developed a numerical model of a tsunami inundation and utilized remote sensing technologies to determine the detailed features of tsunami disaster that killed more than 1,25,000 people (see Asia Site Reports on the International Assessment of Research and Development in Simulation-Based Engineering and Science website, http://www.wtec org/private/sbes/. The required password is available from WTEC upon request). The numerical model predicts the local hydrodynamic features of the tsunami, such as inundation depth and current velocity, based on high-accuracy and high-resolution bathymetry/topography grid. The numerical simulations of the model are validated by the observed records and measured sea levels. Also, high-resolution optical satellite imagery provides detailed information of tsunami-affected area. In recent years, Koshimura is expanding the model's capabilities to comprehend the impact of major tsunami disaster by integration of numerical modeling and remote sensing technologies. Along this line, he integrated his numerical model of tsunami inundation and post-tsunami disaster information obtained from satellite imagery in order to determine the relationship between hydrodynamic features of tsunami inundation flow and damage levels or damage probabilities in the form of fragility functions; development of fragility functions are focused on two cases. One is the fragility function of house damage, which is useful in estimating quantitatively an important aspect of the tsunami's societal impact. The other is the fragility of the mangrove forest (Koshimura and Yanagisawa, 2007). This is important in determining quantitatively the usefulness of the coastal forest to mitigate tsunami disasters.

4.3.2.3 *Osaka University Department of Management of Industry and Technology*

Professor Masaru Zako made a presentation to WTEC team at Tokyo University focusing on development of new simulation capabilities in energy systems, disaster modeling and damage evaluation, life sciences, nanotechnology, and urban environments is impressive (see Asia Site Reports on the International Assessment of Research and Development in Simulation-Based Engineering and Science website, http://www.wtec.org/private/sbes/. The required password is available from WTEC upon request). Professor Zako and his group carried out an extensive study where he uses simulation techniques to deal with hazardous gas dispersion. The aim of their research team is the development of a simulation capability for disaster propagation in chemical plants considering the effects of tank fire and wind speed and direction. Another key aspect is how to estimate associated damages using the results of simulations in conjunction with information from the Geographical Information System (GIS). Monte Carlo methods of estimating radiation heat transfer were studied, and examples from tank fires and gas explosions in realistic scenarios were discussed. The gas originates from leakage from an LNG tank, evaporates and disperses into the air. The dispersion is analytically simulated and traced on a GIS map to predict the regions of contamination. Wireless transmission of real-time wind speed and direction data will make prediction more accurate (Shinozuka *et al.*, 2007b). It is clear that long-term prediction of such events remains a very difficult task. While this study is in nature the same as the study of vehicular emissions of CO_2 gas, the problem simulated here, which can occur following a severe earthquake as the initiating event, can be even more devastating.

4.3.3 *Energy Related Modeling*

4.3.3.1 *Toyota Central R&D Labs Inc. (TCRDL), Japan*

The depth and breadth of SBES activities at TCRDL are impressive (see Asia Site Reports on the International Assessment of Research and Development in Simulation-Based Engineering and Science website, http://www.wtec.org/private/sbes/. The required password is available from WTEC upon request). The role of SBES at the laboratory is in the development and application of simulation tools to aid in understanding phenomena that are observed by experimentalists at the facility. The researchers at TCRDL identified several grand challenge problems for future SBES activities. The most important one is to perform first-principles molecular dynamics simulations of fuel cells to understand the mechanisms of hydrogen atom transport within those systems. The interaction between government and corporate facilities in SBES appears to be beneficial to both parties in Japan.

4.3.3.2 *Fuel Cell Laboratory, Nissan Research Center 1, Japan*

The Nissan fuel-cell research group is mostly doing small simulations on workstations (see Asia Site Reports on the International Assessment of Research and Development in Simulation-Based Engineering and Science website, http://www.wtec.org/private/sbes/. The required password is available from WTEC upon request). Their main target is the membrane-electrode assembly. They are planning a substantial future project: cost reduction and lifetime extension of the Pt catalyst, and substitution of Pt by a cheaper catalyst. They are considering an application of computational chemistry. Models would include rate equations, molecular dynamics, Monte Carlo transport, and dissipative particle dynamics. They plan to collaborate with an outside party to tackle a series of problems: (1) calculate performance of catalysts in a simple electrode reaction and compare results with experimental results, (2) apply computational chemistry to the more complex cathode reaction, and (3) try to predict performance of alternative catalysts, including non-Pt.

4.3.3.3 *Central Research Institute of Electric Power Industry (CRIEPI), Japan*

CRIEPI has an impressive array of simulation and experimental capabilities to address the most challenging problems facing the Japanese power industry (see Asia Site Reports on the International Assessment of Research and Development in Simulation-Based Engineering and Science website, http://www.wtec.org/private/sbes/. The required password is available from WTEC upon request). Its high-end simulation capabilities in materials science and global climate change are on par with the premier simulation institutions around the world. The institution's mission focus enables its researchers to build strongly coupled simulation and experimental campaigns enabling them to validate their simulation results and build confidence in predictions. The close coordination of simulation and experiment should serve as an example to the rest of the world.

The CRIEPI researchers, in collaboration with NCAR (National Center for Atmospheric Research) and Earth Simulator staff, led an international team in performing unprecedented global climate simulations on the Earth Simulator that has influenced international policy on global climate change. The result of the effort was the ability to perform multiple 400-year predictions of future temperature and sea level changes with different CO_2 emissions profiles. The simulations required about six months of wall clock time on the Earth Simulator for four different scenarios to be completed. A challenging high-resolution (a grid spacing of 10 km) ocean simulation was also conducted on the Earth Simulator, and eddy-resolved ocean currents were successfully simulated. Since then, the team at CRIEPI has begun performing regional climate simulations of East Asia to assess the regional impacts of global warming, high-resolution weather forecasting simulations, and global models of ocean currents. Several highlights included simulations of a typhoon making landfall in Japan, and predictions of local flooding in areas surrounding Japan's energy infrastructure that required higher resolution than the Japanese Meteorological Agency typically performs.

CRIEPI researchers provide simulation analysis and tools to the Japanese electric power companies in other areas as well. Examples include CRIEPI's Power system Analysis Tool (CPAT) for electric grid stability, CFD analyses of combustion in coal-fired power plants, CFD analysis of transmission line vibration, and seismic analysis of oil tanks. For the most part, CRIEPI employs commercial software in its routine simulations and outsources code development to tailor the commercial tools for problems of interest. However, the simulation tools for the high-end materials science and climate simulations are developed in-house or with a small set of international collaborators. CRIEPI management considers the codes developed in-house to be proprietary; after establishing the proprietary rights the institution shares its codes with the scientific community.

4.4 Research Activities in Europe

4.4.1 *System Reliability of Electric Power Network*

4.4.1.1 *Union for the Coordination of Transmission of Electricity, Brussels, Belgium*

While this subject matter is of vital importance in Europe as in North America and Asia, it is not strongly addressed in Europe from the view point of system failures due to natural and man-made hazards. In North

America and Asia, severe earthquakes, hurricanes and Typhoons occur frequently. As a result, it appears, the general public, utility companies and regulatory agencies are much more risk conscious and promote the research on this subject. However, the "Union for the Co-ordination of Transmission of Electricity" (UCTE), the association of transmission system operators in continental Europe provides a reliable market base by efficient and secure electric "power highways". Fifty years of joint activities laid the basis for a leading position in the world which the UCTE holds in terms of the quality of synchronous operation of interconnected power systems. Their common objective is to guarantee the security of operation of the interconnected power system. Through the networks of the UCTE, about 450 million people are supplied with electric energy; annual electricity consumption totals approx. 2300 TWh (www.ucte.org). UCTE collected past sequences of black-out events that can be archived to develop scenario of black-out events for future system risk assessment.

4.4.2 *Energy-Related Simulation Research*

There are many interesting studies on harvesting wind and solar energy and integrating the harvested energy into utility power grids. Some of them, which utilize simulation technologies, are selected here from Renewable Energy (Elsevier). These studies are carried out by individual researchers throughout the world demonstrating a strong interest on the subject matter by grass root of the research community. Typical examples of the studies related to wind power are listed below in 1–4 and 5–8 for solar power.

(1) Identification of the optimum locations of wind towers within a wind park using Monte Carlo Simulation (Greece *et al.*, *Renewable Energy* **33** (2008) 1455–1460).
(2) Development of the analytical model to simulate the real-world wind power generating system (France, Ph. Delarue *et al.*, *Renewable Energy* **28** (2003) 1169–1185).
(3) Identification of the most efficient wind power generator by altering its configuration (such as blade type, radius, # of blades, rotational speed of rotor) (Colombia, Mejia *et al.*, *Renewable Energy* **31** (2006) 383–399).
(4) Calculation of efficiency of wind power systems connected to utility grids (Egypt, Tamaly and Mohammed, 2004, IEEE, http://ieeexplore.ieee.org/iel5/9457/30014/01374624.pdf?arnumber=1374624).

(5) Simulation of off-grid generation options for remote villages in Cameroon (Nfah *et al.*, *Renewable Energy* **33** (2008) 1064–1072).
(6) Numerical simulation of the solar chimney power plant systems coupled with turbine in China (Tingzhen *et al.*, *Renewable Energy* **33** (2008) 897–905).
(7) Use of TRNSYS for modeling and simulation of a hybrid PV–thermal solar system for Cyprus (Soteris A. Kalogirou, *Renewable Energy* **23** (2001) 247–260).
(8) Two-dimensional numerical simulations of high efficiency silicon solar cells in Australia (Heiser *et al.*, *Simulation of Semiconductor Devices and Processes* **5** (1993) 389–392).

4.4.2.1 *Technical University of Denmark (DTU), Wind Engineering, Department of Mechanical Engineering (MEK), Denmark*

The number of kW of wind power generated in Denmark per 1000 inhabitants exceeds 570 and is by far the highest in Europe and indeed the world. Spain and Germany follow with 340 and 270 (kW/1000 inhabitants), respectively. By the end of January 2007, there were 5267 turbines installed in Denmark, with a total power of 3135 MW; the total wind power in European Union countries during that period was 56,535 MW with Germany leading at 22,247 MW. The total installed power and the numbers of turbines in Denmark had a continually increase until 2002.

Research on wind energy in Denmark has been taking place for over 30 years, and DTU researchers in collaboration with those of its partner institution Risø have been the leaders in this field (see Europe Site Reports on the International Assessment of Research and Development in Simulation-Based Engineering and Science website, http://www.wtec.org/private/sbes/. The required password is available from WTEC upon request). The research approaches employed at DTU are not only of the fundamental type (e.g., large-eddy simulations and sophisticated grid techniques for moving meshes) but also of practical use, e.g., the popular aeroelastic code FLEX that uses lumped modeling for wind energy prediction. DTU also maintains a unique data bank of international meteorological and topographical data. In addition, the education activities at DTU are unique and impressive. It is one of the very few places in the world to offer both MSc and PhD degrees on wind engineering, and its international MSc program attracts students from around the world, including the United States, where there are no such programs currently.

4.4.2.2 *Institut Français du Pétrole (IFP), France*

IFP's research activities in SBES demonstrate that the institute is deeply grounded in the fundamental mathematical, computational, science, and engineering concepts that form the foundation of SBES projects in the service of technological advances (see Europe Site Reports on the International Assessment of Research and Development in Simulation-Based Engineering and Science website, http://www.wtec.org/private/sbes/. The required password is available from WTEC upon request). The coupling of this sound foundation with the institute's significant and ongoing contributions to the solution of "real world" problems and challenges of energy, environment, and transportation form an excellent showcase for the present capabilities of SBES and its even greater future potential. The IFP research is more applied than that typically found in universities, and its time horizon starting from fundamental science is of longer duration than that typically found in industry development projects. Use of SBES to develop thermophysical properties tables of complex fuel mixtures (biodiesels) at high pressures is a good example of precompetitive research and a platform that benefits the entire automotive industry and all bio-fuel industry participants. The strategy of developing an extensive intellectual property (IP) portfolio for the refining of heavy crude with the view of a refinery feedstock mix shifting in that direction (e.g., increasing share of transportation fuels originating from tar sands) is another illustration of the long time horizon of the IFP. IFP's scientific R&D activities are driven by the following five complementary strategic priorities: Extended Reserves, Clean Refining, Fuel-Efficient Vehicles, Diversified Fuels, and Controlled CO_2.

4.4.2.3 *Science and Technology Facilities Council (STFC) Daresbury Laboratory (DL), Warrington, United Kingdom*

The research group at STFC focuses on the impact of SBES materials research in the energy field, e.g., fuel cells, hydrogen storage, nuclear containment, photovoltaics, novel electronics/spintronics, and solid state battery electrodes (see Europe Site Reports on the International Assessment of Research and Development in Simulation-Based Engineering and Science website, http://www.wtec.org/private/sbes/. The required password is available from WTEC upon request). The overarching and unifying theme is that the design space for the new materials in each of these applications is so large that experiments alone are unlikely to lead to the optimal (or even adequate) materials design, and that with recent and expected advances in algorithms and HPC hardware, the SBES approach is entering the "sweet spot" with respect to applicability and relevance for bridging the length and time scales encountered in foundational scientific predictions for structure & composition, thermodynamics of phases, reaction kinetics & dynamics, and electronic structure/correlation. Research at STFC is timely in the context of solving the looming energy challenges in the post-fossil-fuel era.

4.4.3 *Modeling Civil Infrastructure Systems Resilience and Sustainability*

4.4.3.1 *University of Oxford, Oxford, United Kingdom*

Led by Prof. Allstair Borthwick, the research group at University of Oxford carries out SBE&S research on areas such as river co-sustainability, simulation of dyke break, solitary waves, urban flooding, Tsunami simulation, waves up the beach (see Europe Site Reports on the International Assessment of Research and Development in Simulation-Based Engineering and Science website, http://www.wtec.org/private/sbes/. The required password is available from WTEC upon request).

4.5 Conclusions

It is quite clear that further research is urgently pursued to discover and develop new generation of fuels, alternative to fossil fuel. At the same time, next generation engines must also be developed that can function efficiently with these next generation fuels.

In the United States, energy consumption for buildings and transportation related activities, consumes almost 75% of the energy supply. It is more than prudent then to pursue the search to conserve the energy for construction and maintenance of these and other systems so that the built environment including such systems must be made sustainable. New materials, method of construction or maintenance, novel design guidelines, and the like must all be examined from energy conservation point of view.

Nuclear energy is recognized as the major source for clean energy. Large investments and rapid progress in SBE&S-driven materials design and safety analysis for nuclear waste containment are needed. In relation to alternative source of energy, the energy density is lower but tax incentives for green sources are driving extensive efforts in wind, solar, tidal and geothermal energy. However, better connection technology of these alternative sources of power to the existing power grid remains to be further improved.

In the field of power networks, significant research efforts in simulation techniques to evaluate performance of large-scale and spatially distributed systems such as power transmission systems that are subject to highly uncertain natural hazards (earthquake, hurricanes, and flood) have been given.

Simulation-based engineering is vital for advancing the physical infrastructure of energy, energy production and energy for transportation. It is also an opportunity for creating smart infrastructure. The scarcity of appropriately trained students in the U.S. (relative to Europe and Asia) is a bottleneck to progress.

Overall, U.S. is ahead in modeling of large-scale infrastructure systems, but the gap is closing with Asia in particular. France is ahead in nuclear energy and related research (e.g., containment). Research related to oil production and fossil-fuel supply chain is traditionally led by U.S. Significant research activity is currently taking place in Europe, including new funding models and leveraging strengths in development of community codes. In sustainability front, currently, all regions recognize the need to increase emphasis in this area and research activities are on going.

References

Billinton, R. and Allan, R. N. (1988). *Reliability Assessment of Large Electric Power Systems* (Kluwer Academic Publishers, Boston/Dordrecht/ Lancaster).

Billinton, R. and Li, W. (1994). *Reliability Assessment of Electrical Power Systems using Monte Carlo Methods* (Plenum Press, New York).

Brown, R. E. (2002). *Electric Power Distribution Reliability* (Marcel Dekker, New York).

Endrenyi, J. (1978). *Reliability Modeling in Electric Power Systems* (Wiley, New York).

Gheorghe, A. V., Masera, M., Weijnen, M. and De Vries, L. (2006). *Critical Infrastructures at Risk: Securing the European Electric Power System* (Springer, Dordrecht).

Koshimura, S., Oie, T., Yanagisawa, H. and Imamura, F. (2006). Vulnerability estimation in Banda Aceh using the tsunami numerical model and the post-tsunami survey data, *Proc. 4th Intl. Wkp. Remote Sensing for Disaster Response* (MCEER, New York), http://www.arct.cam.ac.uk/curbe/ 4thInt_workshop.html.

Koshimura, S. and Yanagisawa, H. (2007). Developing fragility functions for tsunami damage estimation using the numerical model and satellite imagery, *Proc. 5th Intl. Wkp. Remote Sensing Applications to Natural Hazards* (MCEER, New York), http://www.gwu.edu/~spi/remotesensing.html.

Na, U. J., Ray-Chaudhuri, S. and Shinozuka, M. (2008). Probabilistic assessment for seismic performance of port structures, *Soil Dynamics Earthquake Eng.*, 28(2), 147–158.

Sanders, B. and Salcedo, F. (2006). Private communication.

Shinozuka, M. (2009). Chapter 3 of the current report.

Shinozuka, M. and Deodatis, G. (1996). Simulation of multi-dimensional Gaussian stochastic fields by spectral representation, *App. Mech. Rev.*, 49(1), 29–53.

Shinozuka, M., Dong, X., Chen, T. C. and Jin, X. (2007a). Seismic performance of electric transmission network under component failures, *J. Earthquake Eng. Structural Dynamics*, 36(2), 227–244.

Shinozuka, M., Karmakar, D., Zako, M., Kurashiki, T. and Fumita, M. (2007b). GIS-based hazardous gas dispersion, simulations and analysis, *Proc. 2nd Intl. Conf. Disaster Reduction* (USA).

Yassin, M. F., Kato, S., Ooka, R., Takahashi, T. and Kouno, R. (2005). Field and wind-tunnel study of pollutant dispersion in a built-up area under various meteorological conditions, *J. Wind Eng. Industrial Aerodynamics*, 93, 361–382.

Zerriffi, H., Dowlatabadi, H. and Farrell, A. (2007). Incorporating stress in electric power systems reliability models, *Energy Policy*, 35(1), 61–75.

Zhou, Y., Banerjee, S. and Shinozuka, M. (2007). Socio-economic effect of seismic retrofit of bridges for highway transportation networks: a pilot study, *J. Struct. Infrastructure Eng.*, accepted.

Chapter 5

NEXT-GENERATION ARCHITECTURES AND ALGORITHMS

George Em Karniadakis

"Gordon Bell's Prize has outpaced Gordon Moore's Law..."

–David Keyes

5.1 Introduction

One of the main recommendations of the NSF SBES report (Oden *et al.*, 2006) is that "investment in research in the core disciplines of science and engineering at the heart of SBES applications should be balanced with investment in the development of algorithms and computational procedures for dynamic multiscale, multiphysical applications." This is particularly important at the present time when rapid advances in computer hardware are taking place and new mathematical algorithms and software are required. This was also stressed in the 1999 report of the President's Information Technology Advisory Committee (PITAC) to the President, which stated, "...information technology will transform science, medicine, commerce, government, and education. These transformations will require new algorithms, new tools, and new ways of using computers..." (Joy and Kennedy, 1999, p. 32), and also, "...Work on system software and algorithms will permit the effective exploitation of parallelism and will contribute to the efficient use of memory hierarchies... Without effective building blocks for software, algorithms, and libraries, high-end hardware is an expensive, underutilized technology..." (Joy and Kennedy, 1999, p. 51).

Recently, in June 2008 and two years earlier than anticipated in the PITAC report, the first ever petaflop benchmark run of Linpack was demonstrated on the Los Alamos parallel computer *Roadrunner*; many more petaflop systems are expected to be operational in the next three years. At the same time, computer designers are developing the first concepts of an exascale (10^{18} ops/sec) computer to be available by 2017 based on more than 10 million processing elements (cores)! While the proponents of these systems that will deliver unique *capability computing* envision new "transformational" and "disruptive" science, it is appreciated by all that the mathematical and system software will play a crucial role for exploiting this capability. The algorithms and computational procedures that the SBES community develops have to take into account these exciting hardware developments, and hence, investments in algorithmic developments and mathematical software should be analogous to investments in hardware.

In 1995 before the teraflop era, a similar argument was raised, namely that "The software for the current generation of 100 gigaflop machines is not adequate to be scaled to a teraflop..." and moreover, "...To address the inadequate state of software productivity, there is a need to develop language systems able to integrate software components that use different paradigms and language dialects" (Sterling *et al.*, 1995). There is, however, a big difference in the technology jump from the gigaflop to teraflop machines compared to what is taking place today in the leap to petaflop and exaflop machines as we approach the limits of transistor technology. While teraflop technology relied primarily on improvements in computer chip speed and a total of less than 1000 processors, the petaflop and exaflop computers will be based mainly on *concurrency* involving, respectively, hundreds of thousands and millions of processing elements with the number of cores per processor doubling every 18–24 months. Moreover, petaflop and exaflop systems are much more expensive to acquire and operate, due to high demand on power. For example, a Linpack benchmark run of two hours on the Roadrunner consumes 2.3 MW, which translates into about two barrels of oil. At this cost, even 1% of petaflop or exaflop computing capability is a terrible thing to waste!

Historically, fast algorithms have played a key role in advancing capability computing in SBE&S at the same or faster rate than Moore's law (i.e., the doubling of computer speed every 18 months). An example is the solution of a three-dimensional Poisson equation, required in many different disciplines from computational mechanics to astrophysics: the Gaussian elimination, used primarily in the 1950s, has computational complexity N^7, which was reduced to N^3 using a multigrid method in the 1980s. This reduction in computational complexity translates into a gain of about 100 million in "effective" flops even for the modest problem size $N = 100$.

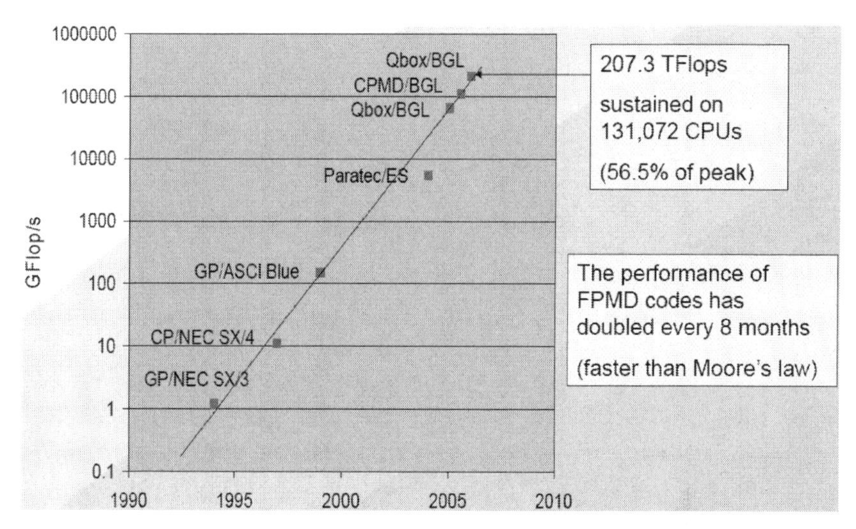

Fig. 5.1. History of the performance of FPMD codes on different computer platforms (courtesy of Francois Gygi, University of California, Davis).

Improved algorithms have resulted in performance gains of several orders of magnitude in many application domains from magnetic fusion to turbulent combustion to molecular dynamics (MD) (Keyes *et al.*, 2003, 2004). Figure 5.1 shows the "effective" Moore's law for a first-principles MD code (FPMD), the QBOX code described by Gygi *et al.* (2005). The slope of the Gflops-over-time curve indicates a doubling of performance every eight months, which is much faster than the aforementioned Moore's law on computer speedup. The main software building blocks of QBOX are basic linear algebra routines (BLAS, ScaLAPACK), and the parallel program is based on MPI (the Message-Passing Interface parallel programming standard). By improving these two areas and using clever decomposition techniques, the QBOX developers have achieved better than 50% of the peak speed of the IBM Blue Gene/L.

The same argument can be made by examining the various application codes that have won the prestigious Bell prize for supercomputing performance. It is clear, then, that Gordon Bell's prize outpaces Gordon Moore's law, due to clever algorithmic and software advances. In particular, despite the diverse pool of applications of the Gordon Bell winners — varying from PDE solvers, to integral equations, to MD simulations — the core solvers are common and are based on the fundamental algorithms of solvers for linear systems, graph partitioning, domain decomposition, high-order discretization, and so forth. Improving these commonly used algorithms and solvers would benefit the broader SBES community.

5.2 High-Performance Computing Around the World

Examining the history of the Top500 list (http://www.top500.org) of supercomputer sites over the last two decades reveals that the United States has been dominant in the world of high-end computing, having about 58% of the "Top500" supercomputers. In the 31st edition of the Top 500 list released in November 2008, the United States had 290 computers in this list while the U.K. had 46, Germany had 25, and Japan had 17 supercomputers. Switzerland had only four supercomputers in the Top 500 list, but if we normalize this number by millions of inhabitants the ratio is 0.53, which is more than half the normalized U.S. ratio of 0.94! It is notable that as of November 2008, three out of the top five systems in the Top 500 list were non-U.S., and Europe has doubled its number of supercomputing systems in the list in the last few years.

5.2.1 *High-Performance Computing in the United States*

In early June of 2008, the world's fastest supercomputer, Roadrunner, which performed the first sustained *one petaflop* speed on the Linpack benchmark, was unveiled by IBM for use at the Los Alamos National Laboratory. Its performance is twice as fast as that of the previous number one supercomputer, IBM's Blue Gene system at Lawrence Livermore National Laboratory. Roadrunner cost about $100 million and consists of 6948 dual-core Opteron computer chips and 12,960 cell engines; its InfiniBand interconnecting system uses 57 miles of fiber optics.

There are many more systems of the petaflop scale on the horizon both in the United States. For example, a Cray XT (1.38 petaflop/s; more than 150,000 cores) was installed in late 2008 at the Oak Ridge National Laboratory, and an IBM Blue Gene/P (1 petaflop; up to 884,736 cores) will be installed at Argonne National Laboratory in early 2009. Several one-petaflop systems will be installed at the supercomputing centers supported by the National Science Foundation (NSF) and connected through the TeraGrid (http://www.teragrid.org). The most impressive NSF project of all, due to be completed in 2011, is the Blue Waters system, an IBM system based on Power 7 multicores, with a total of more than 200,000 cores. Its peak speed will exceed 10 petaflops, but the objective is to sustain one petaflop on real applications.

5.2.2 *High-Performance Computing in Japan*

In Japan, there are three systems of about 100 teraflops peak in the universities, and a 10-petaflop system is on the horizon. Specifically, the University of Toko's T2K system is number one in Japan with a peak of 113 Tflop/s. The TSUBAME at the Tokyo Institute of Technology is number two and beating the Earth Simulator. It has a peak speed of 103 teraflops and is a Sun Galaxy 4 system with Opteron dual cores. A 92 teraflops system from Cray was installed at the University of Tsukuba, and a 61 teraflops system from Fujitsu at Kyoto University. The vector machine NEC SX-9 with peak speed of 839 teraflops will be installed at Tohoku University.[1]

The Japanese government puts forward every 5–10 years a basic plan for science and technology (S&T) that also targets supercomputing infrastructure. The second S&T plan resulted in the installation of the successful Earth Simulator that made Japan number one in the supercomputing world for a while. Currently, the third S&T plan is being implemented, targeting the construction and operation of the Next Generation Supercomputer at the RIKEN labs (see Asia Site Reports on the International Assessment of Research and Development in Simulation-Based Engineering and Science website, http://www.wtec.org/private/sbes/. The required password is available from WTEC upon request). A 10-petaflop system at a cost of $1 billion, due to be ready by 2011, this aims to be the world's fastest supercomputer, targeting grand-challenge problems in nanoscale sciences and life sciences. The system's architecture is a heterogeneous computing system with scalar and vector units connected through a front-end unit; all three major Japanese computer companies (NEC, Hitachi, and Fujitsu) are collaborating on this project. The hope is that parallel processing on such a mix of scalar and vector units will endow the programmer with greater flexibility so that large and complex simulations are efficiently performed.

5.2.3 *High-Performance Computing in Europe*

In Europe, Germany leads the supercomputing activities with three systems in the top 50, including the number 11 spot, a Blue Gene/P system located

[1]Several supercomputing centers in Europe that run climate codes, such as the German Weather Service (DWD) and Météo France, are also expected to install this system.

at the Forschungszentrum (FZ) Jülich. This is the largest national research center in Germany and Europe with an annual budget of €360 million; simulation science is targeted as its key technology. Petaflop systems are planned to be installed at the FZ Jülich and at the University of Stuttgart in the time frame 2009–2011.

France has six systems in the top 50 list. Of particular interest is the Tera-10 project at the French Atomic Authority (CEA) that resulted in its obtaining the number 48 spot in the top 500 list; this is the first-ever supercomputer built and designed in Europe, by the company Bull in collaboration with Quadrics. The Tera-100 project, also at CEA, aims to produce a petaflop machine by mid-2010.

The UK has two systems in the top 50. The European Centre for Medium-Range Weather Forecasts, based at the University of Reading, has an 8320-core IBM P6 system at number 21. At number 46, the 11328-core Cray XT4 system (HECToR), based at the University of Edinburgh, provides a general-purpose high-end platform for use by UK academic researchers. In addition, the university has a 2560-core IBM P5 system, which is also available to the UK research community and intended to complement the service provided by HECToR. Both of these machines are also made available through the UK National Grid Service.

A new European initiative called Partnership for Advanced Computing in Europe (PRACE) has been formed based on the infrastructure roadmap outlined in the 2006 report of the European Strategy Forum for Research Infrastructures (ESFRI, 2006). This roadmap involves 15 different countries and aims to install five petascale systems around Europe beginning in 2009 (Tier-0), in addition to national high-performance computing (HPC) facilities and regional centers (Tiers 1 and 2, respectively). The estimated construction cost is €400 million, with running costs estimated at about €100–200 million per year. The overall goal of the PRACE initiative is to prepare a European structure to fund and operate a permanent Tier-0 infrastructure and to promote European presence and competitiveness in HPC. Germany and France appear to be the leading countries.

Recently, several organizations and companies, including Bull, CEA, the German National High Performance Computing Center (HLRS), Intel, and Quadrics, announced the creation of the TALOS alliance (http://www.talos.org/) to accelerate the development in Europe of new-generation HPC solutions for large-scale computing systems. In addition, in 2004 11 leading European national supercomputing centers formed a consortium, DEISA, to operate a continent-wide distributed supercomputing network. Similar to TeraGrid in the United States, the DEISA grid (http://www.deisa.eu) in Europe connects most of Europe's supercomputing centers with a mix of 1-gigabit and 10-gigabit lines.

5.2.4 *High-Performance Computing in China*

In China, The number 10 system in the world is located in Shanghai at the Shanghai Supercomputer Center with a Chinese Dawning Supercomputer based on AMD processors with a peak speed of 233 Tflop/s. A few HPC centers are located at the major universities, each a few teraflops, connected by gigabit-level network. The top three supercomputers are an IBM/Intel cluster at China Petroleum & Chemical Co., an IBM SP Power4 at the China Meteorological Administration, and the DAWNING Opteron Cluster at the Shanghai Supercomputing Center (see Asia Site Reports on the International Assessment of Research and Development in Simulation-Based Engineering and Science website, http://www.wtec.org/private/sbes/. The required password is available from WTEC upon request). There is a strong government commitment to install by the end of 2009 at least one petaflop in *total* performance and to install in 2010–2011 the first single one-petaflop system.

5.2.5 *High-Performance Computing in India*

India has a system at position 13 on the Top 500 list. That computer installed at the Computational Research Labs (http://crlindia.com/index.htm) and funded by Tata & Sons, Inc., delivers 117 Tflops on the Linpack benchmark. It is based on Hewlett-Packard technology and a proprietary interconnect. Overall, the HPC presence in India is currently weak, and researchers are not driven to push their problems to large HPC environments.

5.2.6 *Future Directions for High-Performance Computing*

5.2.6.1 *Looking to Exascale*

During the spring of 2007, researchers at Argonne National Laboratory, Lawrence Berkeley Laboratory, and Oak Ridge National Laboratory held three town hall meetings to chart future directions on exascale computing systems, their hardware technology, software, and algorithms, as well as appropriate applications for such systems. The goal of the exascale initiative

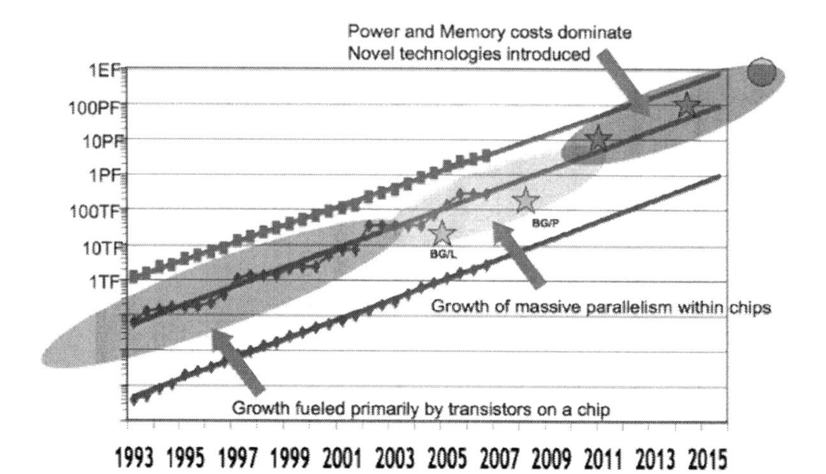

Fig. 5.2. The anticipated path from teraflop to exaflop computing is based on three different types of concurrent technology (courtesy of Rick Stevens, Argonne National Laboratory).

is to achieve exascale performance for a range of SBE&S applications within the next 10 years; this is likely feasible by 2017–2019 (see Fig. 5.2). These higher-performance systems will consist of 10–100 million minicores with chips as dense as 1000 cores per socket; it is expected that clock rates will also accelerate, but slowly. New 3D chip packaging will be required, and the interconnects will be based on large-scale optics. There are many technical challenges to be overcome, including power consumption and leakage currents — large-scale integration and reliability issues related to both hardware and software faults and their detection. A three-step path approach was discussed in the town hall meetings, targeting a 25-petaflops system by 2012, a 300-petaflops system by 2015, and a 1.2-exaflops system by 2019.

5.2.6.2 *Special-Purpose Processors*

The multicores used in the aforementioned petaflop systems are general-purpose processors, but there has been a lot of activity recently to introduce new ways of computing in SBE&S based on special-purpose processors originally built for other applications. For example, Graphic Processing Units (GPUs, e.g., NVIDIA G80) are used in desktop graphics accelerators, Field-Programmable Gate Arrays (FPGAs) are used for digital signal processing, and Sony and IBM cell broadband engines are used for game console and digital content delivery systems. The first petaflop computer, the Roadrunner, used 12,960 IBM Cell chips to accelerate the computation in what is typically termed a *hybrid computing* mode, i.e., mixing standard general-purpose processors with special purpose processors. GPUs, which are ideally suited for applications with fine-grained parallelism, are already yielding speed-ups of factors of 2–200 depending upon the application (source: www.nvidia.com/object/cuda_home.html).

FPGAs, on the other hand, allow for high-performance *reconfigurable computing* (HPRC), i.e., that enables programmers to make substantial changes to the data path itself in addition to being able to control flow. Reconfigurable hardware is tailored to perform a specific task; once the task is completed, the hardware is adjusted to perform other tasks. The main processor controls the behavior of the reconfigurable hardware, as depicted in Fig. 5.3. HPRC can provide higher sustained performance, and it requires less power compared to microprocessor based systems — a critical issue in HPC.

A research group at the National Center for Supercomputing Applications (NCSA) at the University of Illinois at Urbana-Champaign has been evaluating the porting of application codes (cosmology, molecular dynamics, etc.) on FPGA-based systems as well as on other special-purpose processors, obtaining significant speedups. Similarly, several U.S. research groups have been performing large-scale simulations on GPU clusters, including full 3D Lattice Boltzmann simulations of urban dispersion modeling in New York City by a group in Stony Brook University and

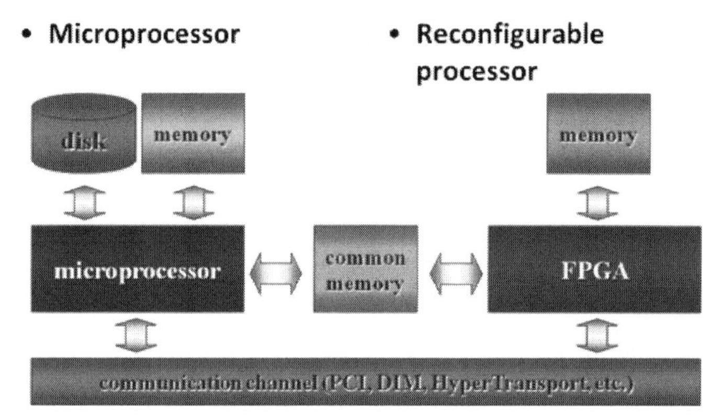

Fig. 5.3. Overview of a high-performance reconfigurable system (courtesy of Volodymyr Kindratenko, National Center for Supercomputing Applications, University of Illinois at Urbana-Champaign).

bioinformatics and protein folding by a group at Stanford University. GPUs can be used as a single cluster or for hybrid computing, i.e., as accelerators in conjunction with general-purpose processors. GPUs allow the *data parallel* programming model, because they include more transistors than CPUs devoted to data processing rather than to data caching and flow control. The NVIDIA Tesla GPU was designed specifically for HPC applications.

5.3 New Programming Languages

Most of the new petascale systems are built on multicores with complex memory hierarchies. From the programming standpoint, these new large multicore systems present another major disruptive technology — a challenge even greater than cluster computing and message passing. According to John Hennessy of Stanford University, this is "the biggest problem Computer Science has ever faced." Simple porting of highly scalable codes like the QBOX or any computational mechanics codes will not lead to any reasonable efficiency, and it carries the danger of largely underutilizing these very expensive petaflop systems, as the PITAC report suggested. Researchers will need to rewrite their application codes and rethink carefully their corresponding algorithms and software. Numerical libraries such as ScaLAPACK will have to undergo major changes to accommodate multicore and multithread computing. The complexity that this involves can be illustrated with a simple example: consider a system with 128 cores per socket, with 32 sockets per node, and with 128 nodes per system for a total of 524,288 cores. Let us also assume that the system has four threads of execution per core. Hence, the total number of threads that a researcher has to handle is two million, a large number of threads indeed!

Management of so many threads is an almost impossible task, even for very experienced programmers, and therefore new programming languages are required to deal with this enormous multithreading complexity. New languages from the High Productivity Computing Systems (HPCS) program of the U.S. Defense Advanced Research Projects Agency (DARPA) point the way toward the next-generation programming environment on petaflop computers. Specifically, Partitioned Global Address Space (PGAS) languages for the SPMD (single program multiple data) model, such as Unified Parallel C (UPC), co-Array Fortran, and Titanium, and dynamic languages (X10/IBM, Fortress/Sun, Chapel/Cray), offer many advantages for programming in the new environments. They support both private and shared data, distributed data structures, one-sided memory communications, synchronization, and collective communications. PGAS languages are a good fit not only to shared memory computers but also to hybrid shared/distributed architectures, and they have locality that may be important when dealing with more than 100 cores per chip.

UPC, in particular, is expressive enough to allow programmers to hand-optimize. In terms of the dynamic languages, X10 emphasizes parallel safety, whereas Fortress and Chapel emphasize expressivity. Fortress uses a math-like representation; X10 and Chapel employ a more traditional programming language front-end. Before the HPC community is likely to adopt these new languages, however, there are many challenges to be addressed, including interoperability with the existing languages, scalable memory models, and parallel I/O procedures.

5.4 The Scalability Bottleneck

The main obstacle to realizing petaflop speeds on multi-petaflop-scale systems remains the scalability of algorithms for large-scale problems. For example, even the scalable code QBOX (see Fig. 5.1) will have a very poor efficiency on a petaflop system when the size of the problem N is large. The reason is very simple: QBOX and many other FPMD codes employ $O(N^3)$ complexity algorithms, which will not scale well for large values of N; hence, we need to develop $O(N)$ algorithms for petaflop computing. Other research teams (e.g., see the site report for the University of Zurich in Europe Site Reports on the International Assessment of Research and Development in Simulation-Based Engineering and Science website, http://www.wtec.org/private/sbes/. The required password is available from WTEC upon request) attempt to literally "split" the atom, that is, to develop new domain composition techniques where parts of the calculation for the same atom are distributed on different processors. For codes that involve solution of a linear system, e.g., solutions of the Poisson equation, the bottleneck is the linear system solver. For example, a solver with scaling $O(N^{3/2})$ takes more than 90% of the time, whereas the remaining 10% is spent on the $O(N)$ complexity part of the code. Hence, the entire computation is almost "all-solver" when the size of the problem increases. It is true that most application groups have not felt this bottleneck — yet, but they will as they start utilizing more than 1000 processors and scaling up their problems.

Scalable usually implies optimal, which for a numerical analyst implies that the floating point operations grow approximately linearly with the size of the problem N. This, in turn, means that for iterative solvers the number of iterations should be constant or almost constant, e.g., they should be bounded "polylogarithmically." The number of iterations is strongly dependent on the condition number of the matrix in the linear system unless effective preconditioners are used to weaken this dependency. Unfortunately, some of the best preconditioners are not parallelizable; for example, the often-used geometric multigrid is fundamentally a serial algorithm, unlike the ineffective but embarrassingly parallel Jacobi preconditioner. However, progress can be made by reformulating existing algorithms to be aware of the topology of the computer architecture.

There are two main challenges for scalable computing as we examine *asymptotic* algorithmic scalability. The first one is expressed by *Amdhal's law*, limiting the maximum speedup by the fraction of the purely sequential work, especially for applications with fixed problem size (strong scaling), e.g., in doubling the number of processors for a fixed problem. The second one is *Little's law* from queuing theory, which, translated into memory bottlenecks, implies that the number of concurrent operations required to avoid waiting on memory is equal to the memory latency in cycles. To deal with these limitations, the simulation scientist has to select carefully the algorithms to be employed, depending on the system architecture. For example, the use of high-order methods increases the memory efficiency and the volume-to-area ratio, hence limiting the required communications. The latter can be overlapped with useful work by splitting operations that use "far away" resources and by employing cross-iteration transformations.

Other approaches may involve development of new algorithms that adapt to memory and concurrency, especially for special-purpose computers, as discussed above. It is also understood from the multigrid example that simple decompositions may lead to large inefficiencies, and similarly, the standard graph partitioning techniques for highly unstructured meshes may be inadequate.

An example of a new type of domain decomposition is shown in Fig. 5.4 for simulating blood flow in the cerebrovasculature, including the Circle of Willis (CoW),[2] the carotids, and other arteries around CoW. The overall

[2]The Circle of Willis (gray structure at the center of Fig. 5.4) is a ring-like arterial network sitting at the base of the brain that functions to evenly distribute oxygen-rich arterial blood throughout the brain.

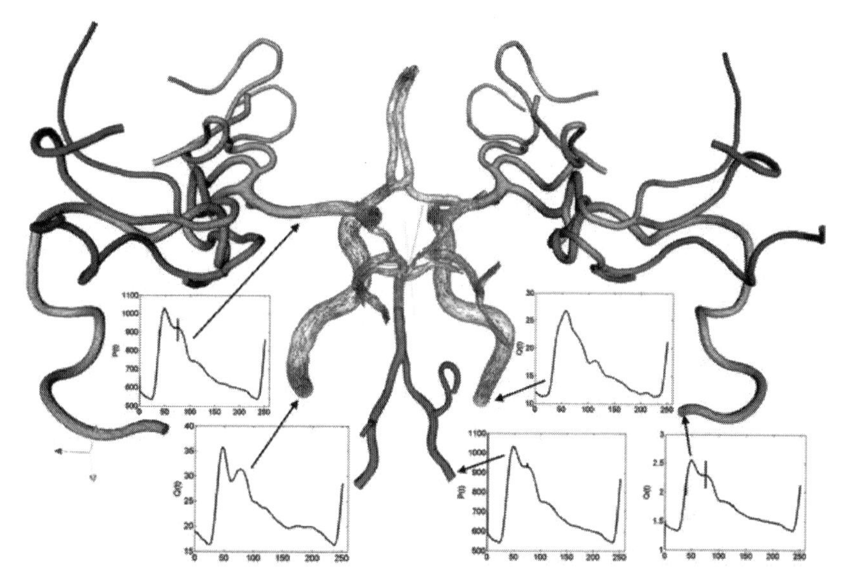

Fig. 5.4. Simulation of blood flow in the human cranial arteries using a new multilayer decomposition technique. Domains of different colors are loosely coupled, whereas within the domain, tighter coupling is maintained. Computations were performed on the Ranger at the Texas Advanced Computing Center (courtesy of Leopold Grinberg, Brown University).

domain is split into several subdomains that are colored differently and assigned to different groups of processors; within each subdomain, the Navier-Stokes equations are solved using the high-order spectral element method, while at the interfaces, proper boundary conditions are employed to ensure continuity of the solution and satisfy the conservation principles. Such a problem-decomposition into subproblems is necessary to avoid the very large condition number of the linear system if the entire domain is discretized directly, but it is also necessary to provide flexibility in the mapping and optimization of work onto groups of processors for each individual subdomain.

5.5 Summary of Findings

In the last decade, large investments have been made for developing and installing new computer architectures that are now breaking the petaflop barrier, and we are looking to an exascale era within the next 10 years. However, recent investments in mathematical algorithms and high-quality mathematical software are lagging behind. It is appreciated by all SBE&S

scientists that the many orders-of-magnitude in speedup required to make significant progress in many disciplines will come from a combination of synergistic advances in hardware, algorithms, and software. Specifically, on the algorithmic front, advances in linear solvers, high-order spatio-temporal discretization, domain decomposition, and adaptivity are required to match the new multicore and special-purpose computer architectures and increase the "effective" performance of these expensive supercomputers.

An example of this synergistic "effective" speedup is offered by a roadmap envisioned by Stephen Jardin of the Princeton Plasma Physics Laboratory and David Keyes of Columbia University on the ITER project, an international research endeavor for fusion energy (Sipics, 2006). To simulate the dynamics of ITER for a typical experimental "shot" over scales of interest would require computers with 10^{24} floating point operations, well beyond the terascale platforms available for production computing today. The Jardin–Keyes roadmap outlines a plan on how to achieve the 12 orders of magnitude required to bridge this gap: three orders of magnitude will be gained through hardware, with both processor speed and increased parallelism contributing equally; the other nine orders of magnitude will come from algorithms (software), namely four orders due to adaptive gridding, three orders due to implicit time-stepping, one order due to high-order elements, and one order due to the reformulation of the governing equations in field-line coordinates. Hence, the main speedup (seven orders of magnitude) will come from better spatio-temporal discretization algorithms that will potentially benefit all SBE&S communities and not just the fusion computational community. Some of these algorithms have already been incorporated in the more advanced parallel codes in the U.S. national laboratories.

Overall, the United States leads both in computer architectures (multicores, special- purpose processors, interconnects) and applied algorithms (e.g., ScaLAPACK, PETSC), but aggressive new initiatives around the world may undermine this position. In particular, the activities of the Department of Energy laboratories have helped to maintain this historic U.S. lead in both hardware and software. However, the picture is not as clear on the development of *theoretical* algorithms; several research groups across Europe (e.g., in Germany, France, and Switzerland) are capitalizing on new "priority programs" and are leading the way in developing fundamentally new work for problems on high dimensions, on $O(N)$ algorithms, and specifically on fast linear solvers related to hierarchical matrices, similar to the pioneering work in Europe on multigrid

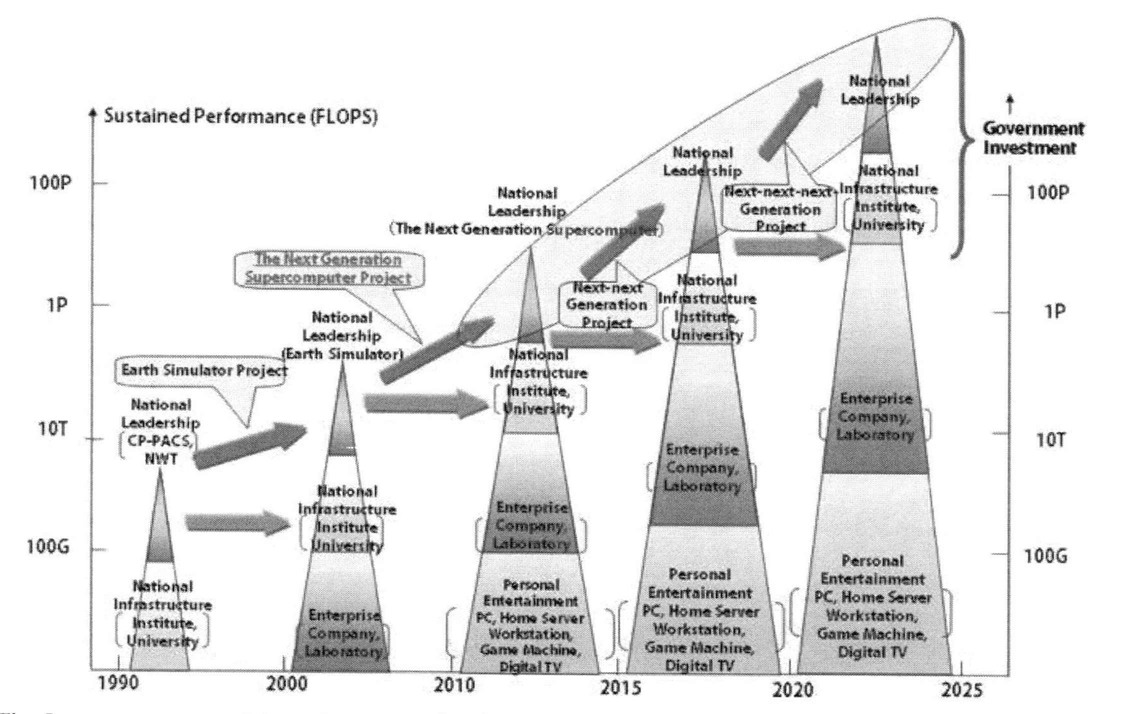

Fig. 5.5. The Japanese government has a long-term plan for sustaining leadership in supercomputing (courtesy of Ryutaro Himeno, RIKEN).

in the 1980s. Initiatives such as the "Blue Brain" at EPFL and "Deep Computing" at IBM-Zurich (see Europe Site Reports on the International Assessment of Research and Development in Simulation-Based Engineering and Science website, http://www.wtec.org/private/sbes/. The required password is available from WTEC upon request) have set ambitious goals far beyond what the U.S. computational science community can target at present.

The Japanese government (through the Ministry of Education, Culture, Sports, Science, and Technology, MEXT) is committed to sustaining supercomputing funding and software activities, as indicated by Fig. 5.5, and is facilitating partnerships and collaborative projects between national labs, academia, and industry, similar to the model of building and operating the Next Generation Supercomputer at RIKEN (see Asia Site Reports on the International Assessment of Research and Development in Simulation-Based Engineering and Science website, http://www.wtec.org/private/sbes/. The required password is available from WTEC upon request). MEXT is, however, emphasizing hardware and application software, and it is less willing to invest on fundamental work in scientific computing or middleware, unlike the European efforts.

It is clear that barriers to progress in developing next-generation architectures and algorithms are increasingly on the theoretical and software side and that the return onto investment is more favorable on the software side. This is manifested by the fact that the "half-time" of hardware is measured in years; the "half-time" of software is measured in decades; whereas the theoretical algorithms transcend time.

Acknowledgements

I would like to acknowledge the help and material provided by our hosts (too many to mention separately) in Japan and Europe. The presentations of Jack Dongarra, Bill Gropp, and David Keyes at the November 2007 WTEC workshop reviewing SBE&S research in the United States have formed the basis for this chapter. I am grateful to Jack Dongarra and David Keyes for correcting the original manuscript. I would also like to acknowledge Rick Stevens for providing material related to the exascale project.

References

Brown, D. L. (2008). *Applied Mathematics at The U.S. Department of Energy: Past, Present and A View to The Future* (Lawrence Livermore National Laboratory, USA), http://siam.org/about/news-siam.php?id=1345.

European Strategy Forum for Research Infrastructures (2006). *European Roadmap for Research Infrastructures: Report 2006* (Office for Official Publications of the European Communities, Luxembourg), ftp://ftp.cordis.europa.eu/pub/esfri/docs/esfri-roadmap-report-26092006_en.pdf.

Gygi, F., Yates, R. K., Lorenz, J., Draeger, E. W., Franchetti, F., Ueberhuber, C. W., de Supinski, B. R., Kral, S., Gunnels, J. A. and Sexton, J. C. (2005). Large-scale first principles molecular dynamics simulations on the BlueGene/L platform using the QBOX code. *Proc. 2005 ACM/IEEE Conf. on Supercomputing* (IEEE Computer Society).

Joy, W. and Kennedy, K. (1999). *Information Technology Research: Investing in Our Future.* (President's Information Technology Advisory Committee, Washington, DC), http://www.nitrd.gov/pitac/report/.

Oden, J. T., Belytschko, T., Hughes, T. J. R., Johnson, C., Keyes, D., Laub, A., Petzold, L., Srolovitz, D. and Yip, S. (2006). *Revolutionizing Engineering Science through Simulation: a Report of the National Science Foundation Blue Ribbon Panel on Simulation-Based Engineering Science* (National Science Foundation, Arlington, VA), http://www.nsf.gov/pubs/reports/sbes_final_report.pdf.

Keyes, D., Colella, P., Dunning, T. H., Jr. and Gropp, W. D. (2003). *A Science-Based Case for Large-Scale Simulation*, vols. 1 and 2 (Office of Science, U.S. Department of Energy, Washington, DC), http://www.pnl.gov/scales/.

Sipics, M. (2006). Taking on the ITER challenge: Scientists look to innovative algorithms, petascale computers, *SIAM News* 39(7).

Sterling, T., Messina, P. and Smith, P. H. (1995). *Enabling Technologies for Petaflops Computing* (MIT Press, Cambridge, MA).

Chapter 6

SOFTWARE DEVELOPMENT

Martin Head-Gordon

6.1 Introduction

The practical applications of computational engineering and science are built around effective utilization of the products resulting from *software development*. These products span a huge range of uses that impact research and development in industries ranging from pharmaceutical and chemical, to semiconductor microelectronics, to advanced materials, to engineering design in automotive and aerospace sectors and many more. The economic significance of these industries is very large — vastly greater than the relatively small companies that typically produce the software itself. Their task is to create integrative tools that bridge basic advances in computer science and the physical and biological sciences from universities with the need for turnkey solutions in engineering design, research and development.

There are fundamental differences between software for computational engineering and science and the software we buy for home or home office. The number of customers is small — for instance there may only be on the order of 100 significant companies in a particular target industry — rather than enormous. At the same time, the simulation problem to be solved — such as perhaps the modeling a catalyst for fuel conversion — involves such complexity and physical realism that the resulting software is often on the order of millions of lines of code, and its development represents research much more than development. Therefore, there is a crucial role for government research support, as part of the overall question of research in simulation-based science and engineering.

There are three broad levels of software relevant to simulations. At the topmost level, discipline-specific applications software is what is employed by a user to solve practical problems, ranging from engineering to physical science to biology. That is what a majority of this chapter will focus on. However, it is important to realize that most applications software also depends upon two further layers of lower-level tools. At the bottom lies the programming environment (what computer language is employed, and what model is used to direct multiple processors), whose development is an applied branch of computer science. Advances in programming paradigms represent an investment in applied computer science that is essentially a mandatory prerequisite for exploiting changes in computer hardware such as massive parallelism or parallelism on a chip. Programming environments tend to lag behind new developments in hardware, and applications software in turn tends to trail behind the provision of appropriate development environments. Finally in between the programming environment and the applications software often lies a crucial middle layer of software — indeed often called middleware — that might include numerical libraries, collections of objects for building applications or interfaces, etc.

Advances in software capabilities have revolutionized many aspects of engineering design and scientific research already — ranging from computer-aided design in the automotive industry to the development of the chlorofluorocarbon replacements that have allowed us to start reversing stratospheric ozone depletion. The significance of these end-use examples is why government support for the development of the next generation capabilities is a good strategic investment — particularly when pursued consistently over the long term. A famous example of simulation software at the applications level that originated with basic research is the Aspen program for process design and optimization in chemical engineering. Now used by virtually all the world's leading chemical and oil companies (as well as many others), AspenTech originated as a spin-off from basic research sponsored by the U.S. government starting in the late 1970s. Similarly, one of the most successful pieces of middleware — the dense linear algebra library called LAPack — is an outgrowth of basic U.S.-government-sponsored research in applied mathematics, and now comprises a package that is ported to and then carefully optimized for essentially all significant computer architectures by the vendors themselves, to encourage subsequent porting of applications software that depends on this middleware.

Those are established historical success stories. The purpose of this chapter is to provide a comparative assessment and discussion of the present American position relative to the rest of the world regarding the software development for simulation-based engineering and science. The discussion and analysis will attempt to address questions such as the following. What is the role of government, if any, in the facilitation of software development for simulation-based engineering and science? If such government role is appropriate, then is the 2008 U.S. level of government investment sufficient to ensure leadership capabilities that will permit us to continue to reap the downstream technological and economic benefits? Is any lack of investment something that is particular to simulation-based software, or is it simply part of larger trends? What are some specific examples of areas where lack of investment has had consequences for software leadership, and what, if any, are the kick-on implications? After a series of sections that attempt to address these issues in the context of activities occurring around the world, and particularly in Europe and Asia, the chapter ends with a short summary of main findings.

6.2 Role of Universities, National Laboratories, and Government

There is worldwide recognition that state-of-the-art simulation software development relies on vibrant activity at the university and national laboratory level from which to obtain new advances — for instance more accurate and realistic simulation models, or new simulation paradigms, or new software engineering advances to enable code development. With globalization, America's basic industrially-sponsored research in computational science and engineering as previously represented by flagships such as Bell Laboratories and IBM's Research Division has declined. While there is still tremendous investment in product-level development, the gap in basic research has had to be made up through increased activities at universities and government laboratories. Furthermore, in science and engineering, software development activities are tightly integrated with innovations in simulation models and algorithms, as well as training. Therefore, a strong university software development community working on simulation method trains the scientists and engineers who will then deploy simulation methods in industry or further develop them in their own research careers. It also produces software that might not otherwise be developed by industry, as shall be discussed later in this section.

Fundamentally, we have found the situation to be broadly similar in America's main Asian and European rivals. The leading industrial sites we were able to visit described their main simulation-based activities as involving the application of tools that were largely developed elsewhere (i.e., in universities and national laboratories). For instance, at the research laboratories of Nissan in Japan, simulation-based modeling is being used as a way of exploring new ideas for improvements in fuel cells, for future vehicles (see site report). A second example was Mitsubishi Chemical (see site report), where simulation software is extensively employed for process simulation, polymer modeling, and materials development problems, as well as others. The diversity of simulation needs relative to the number of staff directly involved makes software development prohibitive. However, Mitsubishi researchers have developed software tools to allow information passing from one program to another (for instance from chemical reaction modeling to flow modeling software), and additionally have contributed to community codes such as AbInit (Gonze *et al.*, 2002). At Toyota Corporation Central Research and Development Laboratories, not only were simulation methods applied to relevant problems, including fuel cell development, aspects of efficient combustion, crash modeling, reduction of engine vibrations, optimizing aerodynamics, but also a number of the key pieces of software were developed in-house.

In Europe, the WTEC panel visited BASF, probably the world's leading chemical company, which has extensive applications-based usage of simulation methods. On the whole, these efforts use codes that are developed elsewhere, although in areas where BASF feels there are critical unmet development needs, or where they feel the company could benefit significantly from advances in simulation capabilities, they follow the strategy of funding extramural research at university groups who can supply such tools (for example in process scheduling), or they participate actively in European Union research networks (for example, BASF coordinates one initiative entitled NanoModel, and also participates in a consortium for the development of theoretical chemistry software). This model of tracking developments in academic research that are of relevance to the company in turn couples to the extensive uses of simulation within BASF. Within China, simulation activity in industrial research is still quite limited with the exception of the oil industry (Sinopec) which we could not visit. Academic researchers whom we talked to felt that there would be greater adoption of simulation methods within industry in the near future, as Chinese industry aims to move up the so-called value chain towards design as well as manufacturing. It is nearly certain that universities will play a crucial enabling role in software development as this transition develops.

Of course the reason that simulation codes are generally not directly developed in industry is because they are only a means towards a vastly more valuable end — the target drug, or fuel cell or aerospace product, which is the real goal of corporate activity. In fact, because the market for simulation codes is so small (perhaps 100 leading chemical companies, perhaps 100 oil companies, perhaps 100 drug companies, etc.), there is only marginal value in a software industry to produce the simulation software directly. The value equation is further tilted by the fact that the simulation software is more often than not directly based on new results from research into simulation models and algorithms. Thus the production of software is inseparable from long-range basic research into models and algorithms. The third (and strongly related) factor disfavoring direct industrial development of simulation software is discussed in more detail in the section on the complexity inherent in state-of-the-art simulation software. This means that the development time can be very long.

For all these reasons, in fields such as materials design, drug design, and chemical processing, where these software factors are at play, software development at universities and national laboratories plays a crucial role. In effect they serve as engines to produce vastly subsidized products for use in industrial research and development — products that for the reasons discussed above would probably not otherwise exist. This transfers direct costs and risk away from the industries that benefit from the use of the resulting software, but would likely not be willing to directly support the real cost of the development of the tools. Of course, in some other fields, such as computer-aided design and computer-aided manufacturing, where the market is larger and the tools are more standardized, this is not so much the case in 2008. However, the closer we are to the frontier application areas of simulation, the greater the role of universities and national laboratories becomes in pioneering the new simulation capabilities and directly providing the software necessary to deploy such capabilities.

In turn, this leads us directly to the need for vigorous government funding of software development for simulation-based research in engineering and science. The situation in this regard will be discussed in more detail in the later section on funding trends. It also leads us directly to the need for effective interactions between universities and industries — such as software companies to commercialize university developments. Going beyond funding, this leads to the question of how efficiently and fairly intellectual property (IP) rights can be obtained for software developments performed at universities to enable commercialization. We heard many comments from both industry and European academics that American universities are at present much more difficult to deal with regarding IP issues. In turn this is leading many companies, even American companies, to deal with foreign universities as more attractive partners for advances in research and development. Of course this situation is not restricted to software development by any means, but in fact applies quite broadly, and is one that American universities should urgently respond to. Whether the future will see foreign universities more aggressively asserting IP rights similar to their American counterparts or whether they will continue to exploit their present competitive advantage is of course an open question.

6.3 Software Life Cycle — Managing Complexity

Nontrivial simulation software for production calculations blends algorithmic and computing advances through typically tens to hundreds of man-years of development and often up to millions of lines of source code. *The resulting software life-cycle usually exceeds the life-span of computing paradigms, and certainly exceeds the length of many funding initiatives.* It is important to recognize this reality, and the associated implication that the greatest rewards are the result of sustained investment in high-quality work. Of course there are exceptions, where well-defined technological developments can be greatly boosted by relatively short-term strategic investments (for instance, the development of new programming models and middleware for the coming massively multicore hardware — see the section below on Emerging Opportunities in Software Development), but these become relatively scarce in the applications disciplines. In this section, we briefly discuss the reasons for the long life-cycle, the consequences for software development in universities around the world, and how trends in funding affect software development activity.

Typically a sophisticated simulation software package that is built from an initially blank slate will take between 5 and 10 years to be mature enough to be released. While there are exceptions that we shall mention later, many such efforts are built with a small nucleus of highly committed scientists and engineers as the initial developers. They may typically be led by a professor and involve a small number of graduate students or postdocs who essentially launch their research careers in the process of developing efficient implementations of novel simulation algorithms (and indeed sometimes developing the algorithms themselves). The fact that new simulation codes are most worth developing when there are new ideas to base them on means that significant effort is involved, including perhaps several false starts and wrong turns. Furthermore, the scale of a working production code and the need to cover all bases — from new discipline-specific algorithms to producing appropriate tools for building inputs and analyzing outputs — means that a large code base is required. If the effort is aiming to produce a commercializable product in the end, then all components above middleware may well need to be constructed from scratch. Even new public domain codes tend to build all pieces as new simply because of advances in software languages, and possibly changes in target computer architectures.

In the section on World Trends in Simulation Software Development, we discuss electronic structure software as a case study. Within this area, two relatively recent "clean sheet" software designs coming from Europe are the ONETEP (http://www.onetep.soton.ac.uk) and Conquest codes (http://www.conquest.ucl.ac.uk) (as new codes, they do not yet qualify as widely used codes, documented there), and serve as examples of this small group paradigm. They were efforts initiated to try to achieve simulation costs that scale only linearly with the number of atoms in the calculation (in contrast to the conventional cubic scaling with particle number for conventional codes), and, as a second goal, be capable of running efficiently on parallel computers. Development of these codes began around 2000, and both are now just reaching the stage where they are usable by simulators beyond the immediate developers. The students and postdocs who have done the hard work of algorithm and code development were not able to publish as many papers as their contemporaries who were working on applications of simulation codes to systems of topical interest in materials science or nanoscience. However, they have contributed significantly to the development of linear scaling methodology, as well as addressing important software engineering aspects such as support for sparse matrices that exploits the particular aspects of electronic structure for efficiency. The key junior figures in both codes have all obtained academic positions in the United Kingdom, which indicates an encouraging degree of recognition of the significance of their work by the U.K. materials science community despite relatively limited ability to publish many scientific papers during the slow process of constructing their programs.

Given the long lead times for development of simulation codes, it is natural that the resulting packages have lifetimes that, in healthy, successful cases, are typically longer than 10 years, and in many cases are measured in decades. The lifetime is related to the complexity of the application, and the variety of functionality that must be supported. New codes that support only limited functionality may find initial success through offering greater performance or a particular capability that is unique. By contrast, mature and established codes continue to find success by providing a greater range of functionality, or even simply a familiar user interface. Furthermore, they are often supported by significant developer communities which provide ongoing enhancements. It is only when their performance becomes uncompetitive and their range of functionality is matched by newer rivals (and ceases to be expanded) that they become eclipsed. This situation is certainly very much the case in many fields. For example, in computational quantum chemistry, the Gaussian and GAMESS programs both have roots stretching back to the early 1970s, so that as of 2008, they are approaching 40 years old. In biomolecular simulations, the first programs to emerge in the early 1980s, CHARMM (see Charmm Online) and Amber (see Amber Online), are still among the leading programs employed as of late 2008.

The consequences for support of software development and for the rate of adoption of new computing paradigms and technologies are interesting to consider. A software development project starting from a clean new design will not be completed during the lifecycle of a typical funding initiative. An example of this type is the NWChem program for computational chemistry that was initiated as a public domain effort as part of the construction project for the Environmental Molecular Science Laboratory at PNNL, running from 1990 to 1995. Shifts in scientific computing paradigms, such as vector processors in the 1980s, and massively parallel computers more recently, will be adopted more gradually than the hardware permits, as software trails in its wake. Codes that are relatively simple will be the first to be ported or rewritten, while complex multimillion line codes will migrate more gradually. The bulk of scientific computing is accordingly performed on established architectures where there is a suitable balance between hardware cost and ease of installation/maintenance, and software availability, as dictated by the factors discussed here that currently determine it.

From the applications software perspective, it is desirable to see improved developer tools that can abstract as many details of the architecture of both present and coming computers into natural constructs of programming languages. This is a clear area of challenge for applied computer scientists, in collaboration with hardware vendors, and application developers. Looking ahead to programming paradigms that improve present languages such as C++ and Fortran 95 by permitting easier abstraction of data and functionality, without unduly compromising performance is one such goal. It may also dovetail with the need to go beyond current tools for controlling parallelism such as OpenMP (primarily for shared memory small scale parallelism) and MPI (for distributed memory parallel machines). The ready availability of robust and usable development tools that support the coming generations of new hardware will in large measure determine how quickly and broadly it is adopted — either at the desktop level or at the supercomputer level. One may also hope that improved software development tools will help to shorten the software development cycle, and therefore more effectively enable the development of next generation applications. This can in part be true, although it must again be stressed that the development of truly new applications is equally bottlenecked by the intellectual barriers to developing the new simulation models and algorithms that define them. It is not clear, for instance, that multimillion-line codes based on dated languages such as Fortran 77, are appreciably handicapped in practice against codes that are developed in more up-to-date languages, at least as measured by their continued usage and evolution. For example, several of the codes being ported for use with the Japanese next generation supercomputer are indeed Fortran 77 programs.

Finally, another aspect of complexity is that with increasing focus on multiscale modeling (see Chapter 9), there is a need for improved interoperability of different software programs. This was a need expressed by practicing simulation scientists and engineers at a number of sites visited by the WTEC panel, such as Mitsubishi in Japan and BASF in Germany. For example, software that models chemistry at the atomistic level should be able to provide inputs to software working on longer-length scales that describes fluid dynamics. Or programs with complementary capabilities should be able to exchange data so that one provides input for the other. Neither of these needs is adequately met in the major applications disciplines at present. This could be met by a targeted initiative in this particular form of middleware.

6.4 Supercomputing Software Versus Software for Midrange Computing

Supercomputers represent the leading edge of hardware development, and correspondingly offer the ultimate potential for applications performance at any given time. With the further passage of time, advances made at this level can, when appropriate, trickle down to benefit future developments in midrange computing software. As discussed in detail in Chapter 5, we are in the petaflop era in 2008, and we expect that by early in the next decade, the world's leading computers will be in the 10 petaflop class. The development of software for these machines poses challenges that are at least equal to those of previous generations of supercomputers. The reason is that the new hardware is dramatically different from those where software development is concentrated today — in particular the extent of parallelism is vastly greater. Indeed the degree of parallelism will be the primary differentiator between midrange computing and supercomputing. Most nonclassified supercomputer software development is directed towards long-term grand-challenge scientific applications of the type described in previous chapters — spanning materials, energy, nanoscience, health and medicine. Some applications with a more immediate focus include oil industry simulations of reservoirs, and turbulent combustion for engine or turbine design. However, given the tremendous cost of supercomputing and the need for customized software, it is only in industries where the results of simulations are most mission-critical where it can be justified.

Japan is investing heavily in software development for the next-generation supercomputer project, an effort running from 2006 to 2012 to replace the Earth Simulator. This software investment is because it would be a tremendous waste of resources to produce the hardware, without a corresponding effort to have software in place for this machine, as soon as possible after it is commissioned. Therefore a set of six scientific software challenges have been identified, three originating primarily from the physics community and three from the chemistry community. The physics-derived software will comprise codes for real-space density functional theory (DFT), dynamical density matrix renormalization group (DMRG), and quantum Monte Carlo (QMC). The chemistry-derived applications will comprise software for molecular dynamics (MD), the reference interaction site model (RISM) approach to solvation, and the fragment molecular orbital (FMO) approach to treating the electronic structure of very large molecules. One particularly interesting aspect of the software development projects is that they are underway now before the exact form of the next generation supercomputer is known. This is to allow time for innovation in the algorithms and models, rather than simply being forced to adapt existing codes to the machine when it is available. Since it is widely recognized (see Chapter 5) that advances in applications performance derive at least as much from improvements in models and algorithms as hardware advances, this is a very reasonable attempt to blend the realities of supercomputer hardware development with those of software development.

A second example of very active supercomputer software development was seen at Karlsruhe Institute of Technology (KIT) in Germany. KIT is home to the Steinbuch Center for Computing (SCC), which, in terms of personnel, is Europe's largest with more than 200 staff members. In contrast to supercomputer centers in the United States, the SCC has a number of professors whose positions are explicitly within the center. One of these positions is in application sciences, and involves the development of a variety of simulation software, with capabilities ranging from modeling hull design for an entry in the America's cup, to the modeling of air flow through the nose. All this software development is within an open source model. Other professors have specializations in numerical methods and aspects of high performance computing research. This embedding of long-term research projects in high-performance computing into a supercomputer center is an interesting alternative to the U.S. system where most activity at supercomputer centers is more closely tied to supporting users.

A third example of supercomputing software development was at the Institut Français du Pétrole in Paris, which performs a wide and deep variety of simulations in support of basic and applied petroleum research (see site report). In areas such as molecular modeling, they employ third-party codes. However, they have an extensive effort in high-performance computing software development, such as computational fluid dynamics codes to enable the design of improved internal combustion engines. New-generation large-eddy simulation (LES) codes specifically enable modeling of fast transients, cold starts, etc. Previous-generation codes have been commercialized and then more broadly distributed. Parallel scaling of research codes was demonstrated to the WTEC panel on up to 4000 processors, consistent with the IFP goal of keeping a high-performance computing capability that is roughly a factor of 10 behind the state of the art.

While supercomputer applications are valuable as discussed above, support for software development targeted specifically at supercomputers must be tempered by the fact that its impact is relatively narrow in the sense that the vast majority of useful simulations in scientific and engineering research is accomplished using midrange (or "capacity") computing, usually on clusters that are controlled by small groups of researchers. Relative to 15 years ago, the most remarkable development in scientific computing is the merging of the standard office personal computer with the scientific workstation — there has been no meaningful distinction for a decade or so. Midrange computing is accomplished with clusters that are essentially collections of personal computers, with an extraordinary reduction in the cost relative to the situation that previously existed when such machines were specialized research tools.

This has contributed to two important world trends, both of which were evident in our study trip. First, due in part to the tremendously decreased cost of the computing resources is an enormous broadening of the user base in many fields. This can be illustrated for computational quantum chemistry in the academic literature by the data shown in Fig. 6.1. It is clear that there has indeed been a tremendous increase in adoption of these simulation methods during the period 1999–2007. This is an example of the usage penetrating far beyond the (relatively small) community of simulation experts and into a much larger community of nonspecialist users. This reflects both the diminished cost of entry and the increasing capability of the methods themselves. While it is not possible to survey many other fields in the same way, we strongly suspect that similar considerations apply.

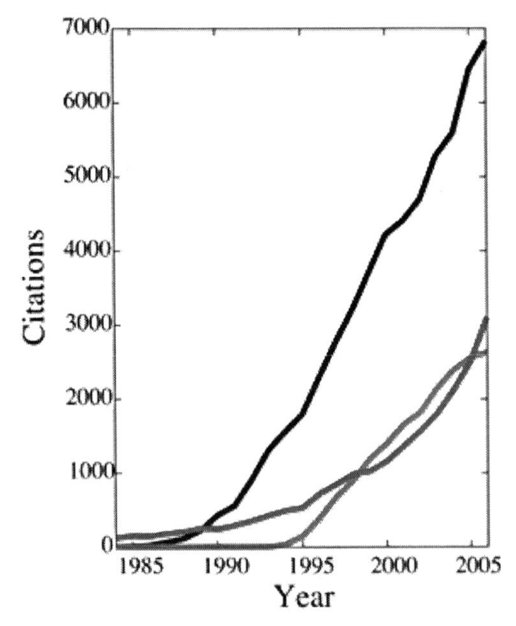

Fig. 6.1. Development of the usage of computational quantum chemistry, as measured by two different metrics. The top curve gives the number of citations to software packages per year, while the lower curves provide the number of citations of particular electronic structure methods (specifically pure and hybrid density functionals). Data are from the Web Science (courtesy of Emilio Artacho). Note it would be interesting to illustrate a modified version that tracks U.S. and worldwide usage separately.

The second trend concerns the expert community itself. Groups that develop and apply simulation methods in the United States traditionally had a large advantage in their computational facilities relative to rivals in either Europe or Asia. This is no longer clearly the case now when the dominant cost in the United States at the midrange level is labor rather than equipment. Indeed the panel found that development resources were broadening across the world and leveling out.

We conclude this section by observing that strategic investments in simulation software development must in the end balance support for advancing the state of the art in supercomputing, with support for innovative developments that affect the far larger communities that rely on midrange computing. As a senior simulation scientist at Karlsruhe, site of Europe's largest supercomputer center (by personnel) remarked: "supercomputing is fine but most science gets along without it."

6.5 World Trends in Simulation Software Development

It is not possible to cover all areas of software applications development, or even a significant fraction, in any meaningful detail in a document of reasonable length. The approach that will be taken here is to first discuss some generalities about the present world situation regarding simulation software development. We shall consider middleware briefly, and then focus primarily on scientific and engineering applications software. We then consider software development in a representative applications area in some detail (electronic structure codes in condensed matter physics and chemistry). This permits some specific illustrations of general trends, which at least in part, show a strong shift away from American leadership in software development, and some of the consequences.

We first discuss some broad generalities. In 2008, the US continues to dominate in the development of system-level tools and middleware, as has been the case virtually since the large-scale computing first became viable in the 1950s and 1960s. In large part this is because of the strength of American leadership in hardware development, and the resulting impetus from hardware manufacturers to spur the development of appropriate software tools that enable effective usage of their platforms. At the supercomputing level, this is often supported by developments at national laboratories and universities because of the relatively small scale of the market. There has been a trend towards moving these tools into the public domain, driven by the success of the open-source development of the Linux operating system and the closely related Gnu software tools. Thus obtaining access to existing system-level software and middleware is now easier and less expensive than it has ever been. Without a doubt this has contributed to a leveling of the playing field for applications development, since there is no clear-cut technical advantage in the tools of software development (or, to a considerable extent in the hardware, as discussed in the midrange computing above).

Leadership in applications software development, therefore, is not at present determined by access to either hardware or development software, and will instead be determined by other factors. Those factors are diverse, but clearly include the strength of the relevant scientific and engineering research communities in the particular discipline, and the incentives or disincentives for software development activity relative to applications of existing software to solve problems of interest. Those incentives and disincentives revolve around what type of work is considered of highest priority for funding, the level of professional recognition and rewards in software development activity relative to targeted applications, and the appropriateness of either activity for graduate student education and training. Indeed, it is possible to even adopt the strategic view that because of the level playing field for hardware and standard middleware, new investments can be focused on trying to drive the applications. In software development for Japan's next generation supercomputer, discussed above, this point of view has been adopted — there will be no concerted development of middleware because of their ability to obtain American-developed middleware, and their desire to compete for software development leadership in applications that impact energy, new materials, and biology.

Our impression is that on the whole, American leadership in applications software development for simulation-based science and engineering has diminished quite noticeably relative to a decade ago (2008 versus 1998). In some areas, U.S.-based applications software development is now clearly trailing competitive efforts in Europe particularly. Japan also has some examples of flagship software development (such as supercomputing applications), but their impact in terms of widespread distribution for mid-range capacity computing has been more limited. One example of such an area is computational condensed matter physics, discussed below. The related area of computational quantum chemistry codes has increasingly strong world-wide competition, but the most widely used commodity codes are still American-developed. In other areas, Europe has also made large strategic investments in software development, such as preparing for the Large Hadron Collider (LHC) experiment, and thus the strategic balance is changing. The panel found that China at present (2008) is not a major player in software development for simulation-based science and engineering. However, the strong push to raise the level of excellence of the leading Chinese universities is leading to a new generation of talented academic scientists who are training talented students in increasingly large numbers, and we should expect a greatly increased Chinese presence in simulation software development in the future. India is also increasingly laying the groundwork for a future presence in simulation software development, due to the combination of a strong presence in commercial software development, a strong academic tradition at its leading universities, and now increasing government investment.

Let us now turn to discuss software for electronic structure calculations as a specific example. Such software solves sophisticated approximations to formally exact but practically intractable quantum mechanics to describe the behavior of electrons in bulk solids (including new materials), surfaces and interfaces (of relevance for catalysis — see the WTEC report on catalysis research by Davis *et al.*, 2009), and molecules and nanoscale materials (relevant for chemical reactions generally). Software development in this area was pioneered in the United States starting in the late 1960s, and, with the passage of time, reached a level of sophistication that began to have a substantial impact on not only academic but also industrial applications. One such example was its use in the 1990s to assist in the design of replacements for chlorofluorocarbons (CFCs) which were by then recognized as contributing to the destruction of the ozone layer. The American physicist, Walter Kohn, who developed the theorems that underlie modern electronic structure software, and the chemist, John Pople, who led the development of the software (the Gaussian series of programs) that first made a major impact on diverse areas of chemical research, in fact shared the Nobel Prize in Chemistry in 1998.

To assess the software development situation in 2008, we shall first discuss the codes that are used to treat extended materials, which are primarily developed in the condensed matter physics community. The first codes in this area were developed by leading American researchers at institutions including Berkeley, Northwestern, Lawrence Livermore, Los Alamos, and many others. Today, however, the codes that are most widely

Table 6.1. Widely distributed physics codes for electronic structure calculations on extended systems (source: WTEC panel).

Program name	Function	Type	Origin
VASP	Plane wave DFT	Commercial	Vienna, Austria
CASTEP	Plane wave DFT	Commercial	Cambridge, United Kingdom
Siesta	Local orbital DFT	Public domain	Barcelona, Spain
CP2K	Plane wave DFT	Public domain	Zurich, Switzerland
Quickstep	Local orbital DFT	Public domain	Zurich, Switzerland
Quantum Espresso	Plane wave DFT	Public domain	Princeton, United States
CRYSTAL	Gaussian orbital DFT/HF	Commercial	Pisa, Italy

used for practical applications come almost entirely from Europe. They are summarized in Table 6.1. These software programs include both commercial programs and public domain programs. They are dominant in application calculations around the world, including the United States, where it is common for leading research groups to employ them. This is despite the fact that scientific leadership in the research that drives new developments in the programs is more evenly balanced than the lopsided situation represented by the national origin of the software packages shown in Table 6.1. Indeed, there are still numerous "in-house" programs in the United States that exist within one or a handful of groups. These codes, it appears, are mostly used to explore new ideas, though in some cases they are also used for "in-house" production calculations. However, as nondistributed codes, they are often lacking either in performance or standard features, and therefore even in groups that have such programs, it is common for them to employ the standard European codes for production calculations.

In Table 6.2, the codes that are widely used to treat molecular systems are shown. They are developed primarily in the computational quantum chemistry community. The situation here is far better balanced between Europe and the United States than in the case for physics, but this author feels that there has been a relative change in the amount of activity in the United States relative to the rest of the world. In straightforward terms, the rest of the world is catching up, and even the U.S.-developed codes often involve substantial and important contributions from groups based in Europe or Japan. The U.S.-based Gaussian program, for example, includes large amounts of software developed in Japan (Kyoto University) as well

Table 6.2. Widely distributed chemistry codes for electronic structure calculations on molecular systems (source: WTEC panel).

Program name	Function	Type	Origin
Gaussian	Gaussian quantum chem.	Commercial	Connecticut, United States
GAMESS	Gaussian quantum chem.	Public domain	Iowa, United States
Q-Chem/Spartan	Gaussian quantum chem.	Commercial	Pittsburgh, United States
NWChem	Gaussian quantum chem.	Public domain	Washington, United States
ADF	Local orbital DFT	Commercial	Amsterdam, Holland
TurboMole	Gaussian quantum chem.	Commercial	Karlsruhe, Germany
MolPro	Gaussian quantum chem.	Commercial	Stuttgart, Germany
MolCAS	Gaussian quantum chem.	Commercial	Lund, Sweden
Dalton	Gaussian quantum chem.	Public domain	Oslo, Norway

as Europe (groups in Italy and England). Likewise, the U.S.-originated Q-Chem program includes key contributions from groups in Australia and Germany. The public domain GAMESS program is not significantly different, with important contributions from Japan (IMS). Therefore, some of the slippage in the U.S. position is disguised by the historical advantage that major codes originated in the United States, but it is occurring nonetheless.

What are the consequences of a loss of U.S. leadership in software development for an applications area such as condensed matter physics electronic structure programs? They begin with issues that may seem essentially benign such as U.S. researchers acquiring the European-sourced codes to enhance their often-leading research on application problems using simulations. This is arguably not so different from the world reliance on U.S.-developed middleware discussed above. However, it is important to remember that the United States restricts the export of high technology items to organizations that may use them to compete militarily against the United States. With software applications dominated by overseas developers, the same effect can and does start to occur in reverse. The WTEC panel was made aware of one specific recent example of this type in the computational condensed matter physics area. The Siesta program, noted in Table 6.1, has an add-on module for calculating transport properties, TranSiesta, with development headquartered in Spain. The request for a license for this program for the AFOSR research base was refused by the TransSiesta developers based on the principle of not contributing to research with potential military applications. Therefore, when simulation software reaches the stage of having the ability to contribute to research on issues regarding national security, the consequences of a loss of national leadership in software development can, and in this case, did, translate to inability to access that software.

There are also consequences for leadership in developing the new ideas for improved algorithms or new simulation models in the future. Since there is a close relationship between producing new ideas and turning them into software (without the latter, the former is not proven useful), there is a real possibility that leadership in software development will synergize with leadership in developing the new simulation paradigms. There is also the question of student training (see Chapter 11 for much more detail on this problem), which can only be effective for software development if research is performed on a state-of-the-art program. Furthermore, students trained primarily to run simulation software are less flexible simulation scientists and engineers than those who are also trained to be able to handle large-scale software development projects.

6.6 Comparative Aspects of Funding for Applications Software Development

We have presented evidence that concerted European and Japanese efforts are leading to a tipping of the competitive balance in some fields, while China is fast developing. Generally, this is a reflection of larger trends — U.S. investment in science and engineering research of most types has been gradually declining in real terms during this decade, while overseas rivals have been enhancing their own spending. In the case of China, the doubling time for the overall science budget has been on the order of every 5 years since around 1990. The most important single thing that could be done is to reinvigorate U.S. research in science and engineering at the single investigator level, based simply on the criterion of excellence. Within the United States, numerous researchers have complained to the panel that in addition to gradual budget cuts, agencies such as the National Science Foundation are overemphasizing so-called broader impacts (communicating research results and educating society at large) relative to scientific excellence, as well as overemphasizing short-term initiatives relative to the long-term health of basic research.

However, in light of the long software development process, there are also specific challenges in obtaining funding for simulation software development within the United States. A typical funding initiative will involve a high level of research support for a period such as 3–5 years, sometimes with the option for a second renewal period subject to achieving satisfactory performance in the first grant. Such a timescale is only marginally suitable for new software construction — though it can be very valuable for enhancing software that is already established. In general, however, short-term U.S. (and overseas) initiatives do not promote the long-term health of innovation-driven software development — rather they tend to encourage novel applications of the tools that already exist. In turn this reduction in applications software development in the United States is leading to reduced production of appropriately trained students which will threaten our long-run competitive position.

Indeed several European investigators commented on the fact that there have been numerous short-term initiatives to build new U.S.-based codes that have then ceased or stagnated when funding ended. They also commented that this situation tended to make U.S.-based scientists somewhat unreliable partners for long-term software development projects. While the academic simulation software development situation is not entirely satisfactory in either Europe or Asia either, it is nonetheless striking that Europe in particular has been able to better sustain a strong tradition of both community and commercial code development. European researchers complained that they were unable to hire personnel specifically for programming tasks, and that obtaining genuinely long-term support for software development was very hard. One exception is that long-term software maintenance of community chemistry codes is supported to a modest degree in the United Kingdom by several long-term Computational Chemistry Projects (CCPs).

One may argue that this state of affairs has at least the merit of encouraging innovation in any new software development project, as that is the primary mechanism by which funding can be obtained. Longer term, the commercialization of software is the most widely used mechanism to obtain funding for code maintenance and development. This approach applies to virtually all of the codes that are commercial which were listed in Tables 6.1 and 6.2, and this is a real incentive to make a code commercial rather than public domain. Users thereby support part of the cost of ongoing development, even though government provides essentially all funding for the addition of new algorithms and models which can be classified as novel research. If government wishes to specifically encourage the development of open source and/or public domain codes, then it is essential that some support for programmers be provided, independent of research funding.

A few words should be added on the type of teams that develop applications-based simulation software. Due to the size and longevity of software codes, they are usually developed by large teams. Generally such teams are based in multiple locations and are quite loosely coupled. Different team members typically work on different pieces of functionality. It is important to note that such teams are often based around a very small core of people at a central site who carry primary responsibility for managing the collaboration and the code. There are countless such examples, such as documented in Tables 6.1 and 6.2. Therefore, funding the development of codes does not necessarily require funding the entire team, but may naturally separate into smaller grants that can be judged based on traditional requirements of innovation and excellence.

Of course these considerations apply to software development as the WTEC panel has seen it in action in 2008. It is possible to imagine that higher levels of funding could support an entire team, including team members such as applied computer scientists or professional programmers who are not often incorporated in typical simulation code teams today. However, this type of support will only be effective if provided for a sufficient period of time to be commensurate with the decade timescale that the WTEC panel has typically observed for the development of state-of-the-art simulation software.

6.7 Emerging Opportunities in Software Development

The specifics of innovation in software development across the vast diversity of active disciplines in simulation-based science and engineering are not predictable. However, the synergistic effects of continuing exponential improvement in hardware performance (including that which can be obtained for an individual researcher, as well as supercomputers) together with improvements and sometimes revolutions in simulation models and algorithms guarantee new opportunities. We could not know that investments in computational chemistry in the 1970s and 1980s would help to save the ozone layer in the 1990s. The energy, materials and health challenges discussed in earlier chapters set the broad context where strategically important benefits are likely to be gained in the future as the result of new simulation capabilities. Largely it will be ensuring dynamism within the application fields in general that will be crucial rather than large centrally organized investments in computational science and engineering. The latter can only serve to partially and unevenly compensate for a general lack of investment in research in the relevant disciplines.

An important theme regarding software development for simulation-based engineering and science is that it should be seen as an opportunity whose time has come to different extents in different fields. Within a discipline it depends upon whether or not we have crossed the point where computational resources permit predictions with useful accuracy. In the example of computational quantum chemistry discussed in the context of Fig. 6.1, it is evident that we have, and therefore further advances in models, algorithms and computer hardware all provide a synergistic effect that enables users to treat larger and more complex molecules. In other fields, such as evolutionary biology, it is reasonably clear that we do not yet have useful computational models at all, and therefore advances in computer capabilities are for the time being entirely irrelevant. Finally in some other fields, the computational requirements are so low that advances in computing do not materially change their ability to do the relevant simulations which were already possible. Thus the automobile industry used to be a major purchaser of supercomputers but this is no longer necessary for CAD/CAM purposes, which can be met with high-performance workstations and midrange computing. By contrast, the oil industry continues to be major users and thus purchasers of high-end computer hardware.

For those fields in which useful simulations are currently computationally limited, new government investments in software development will likely reap future rewards. Many of the opportunities that are highlighted in the chapters on materials, life sciences, energy, multiscale modeling, and algorithms constitute appropriate examples. The algorithms that are contained in software developed for recent generations of hardware contains new capabilities that are only feasible as a result of hardware advances. A good example in the context of electronic structure calculations discussed above is the emerging ability to exploit separations of length scales to obtain algorithms with complexity that scales only linearly with the size of the molecule. In turn, new types of application calculations are enabled by the synergism between improvements in hardware performance and algorithms, together embodied in functional software. The transformational power of new simulation capabilities is greatly magnified when the computing requirements are as inexpensive as commodity personal computers are in 2008.

Therefore, it is worth highlighting the possibilities of the next generations of commodity computers. Advances in graphical processing units (GPUs) have been outpacing advances in general purpose central processing units (CPUs). Indeed the hardware associated with the current generation of video games such as the Microsoft Xbox and the Sony Playstation III, has computational capability that is very similar to state-of-the-art PCs (if one disregards other aspects of system performance such as disk-based IO and memory bandwidth). It is very likely that future commodity processors from companies such as Intel and AMD will integrate general purpose GPUs (GPGPUs) with CPUs. GPUs demand very high degrees of parallelism in order to achieve a reasonable fraction of their peak performance. Accordingly it is likely that programming tools, which are currently a relatively rudimentary stage for GPUs, should be much further developed as a prerequisite for widespread software development activity to exploit the capabilities of these devices for scientific and engineering applications. It is also very likely that other advances in general-purpose commodity computers will be in the direction of far greater parallelism, because as discussed in Chapter 5 advances in computer speed based on increasing clock speed are no longer possible due to physical limitations associated with heat generation.

6.8 Summary of Findings

Modern software for simulation-based engineering and science is sophisticated, tightly coupled to research in simulation models and algorithms, and frequently runs to millions of lines of source code. As a result, the lifespan of a successful program is usually measured in decades, and far surpasses the lifetime of a typical funding initiative or a typical generation of computer hardware. Leadership in many disciplines remains largely in U.S. hands, but in an increasing number of areas it has passed to foreign rivals, with Europe being particularly resurgent in software for midrange computing, and Japan particularly strong on high-end supercomputer applications. Moreover, defense research in the United States has been denied access to foreign-developed software as a matter of principle.

Simulation software is too rich and too diverse to suit a single paradigm for progress. Choices as disparate as software targeting cost-effective computer clusters versus leading edge supercomputers, or public domain software versus commercial software, choices of development tools, etc., are mapped across the vast matrix of applications disciplines. Best outcomes seem to arise from encouraging viable alternatives to competitively co-exist, because progress driven by innovation occurs in a bottom-up fashion. Thus strategic investments should balance the value of supporting the leading edge (supercomputer class) applications against the trailing vortex (midrange computing used by most engineers and scientists).

Fundamentally, the health of simulation software development is inseparable from the health of the applications discipline it is associated with. Therefore, the principal threat to U.S. leadership comes from the steady erosion of support for first-rate, excellence-based single-investigator or small-group research in the United States. A secondary effect that is specific to software development is the distorting effect that long development times and modest numbers of publications have on grant success rates. Within application disciplines, it is important to recognize and reward the value of software development appropriately, in balance with the direct exploration of phenomena. Software development benefits industry and society through providing useful tools too expensive, long-term, and risky to be done as industrial R&D, trains future scientists and engineers to be builders and not just consumers of tools, and helps to advance the simulation state of the art.

Future investments in software development at the applications level are best accomplished as part of re-invigorating U.S. physical and biological sciences generally. More specific investments in simulation software can be justified on the basis of seeking targeted leadership in areas of particular technological significance, as discussed in more detail in chapters on opportunities in new energy, new materials, and the biological sciences.

References

Amber Online, http://ambermd.org/.

Anjyo, K. (1991). Semiglobalization of stochastic spectral synthesis, *Vis. Comput.*, 7, pp. 1–12.

Anjyo, K., Usami, Y. and Kurihara, T. (1992). A simple method for extracting the natural beauty of hair, *Comput. Graphics*, 26(2), 111–120.

Charmm Online, http://www.charmm.org/.

Davis, R. J., Guliants, V. V., Huber, G., Lobo, R. F., Miller, J. T., Neurock, M., Sharma, R. and Thompson, L. (2009). *WTEC Panel Report on International Assessment of Research and Development in Catalysis by Nanostructured Materials* (World Technology Evaluation Center, Inc., Baltimore).

Gonze, X., Beuken, J. M., Caracas, R., Detraux, F., Fuchs, M., Rignanese, G. M., Sindic, L., Verstraete, M., Zerah, G., Jollet, F., Torrent, M., Roy, A., Mikami, M., Ghosez, P., Raty, J. Y. and Allan, D. C. (2002). First-principles computation of material properties: the ABINIT software project, *Comput. Mater. Sci.*, 25, 478–492.

Chapter 7

ENGINEERING SIMULATIONS

Abhijit Deshmukh

7.1 Introduction

The use of simulation and computational models is pervasive in almost every engineering discipline and at every stage in the life cycle of an engineered system. Typical examples of the use of simulation in engineering include manufacturing process modeling, continuum models of bulk transformation processes, structural analysis, finite element models of deformation and failure modes, computational fluid dynamics for turbulence modeling, multiphysics models for engineered products, system dynamics models for kinematics and vibration analysis, modeling and analysis of civil infrastructures, network models for communication and transportation systems, enterprise and supply chain models, and simulation and gaming models for training, situation assessment, and education (PITAC, 2005; Oden *et al.*, 2006).

Simulation and modeling approaches used in engineering can be classified along various dimensions, other than application domains. One classification is based on the use of the simulation models, whether they are used for analysis or synthesis of the system. In most cases, solving the analysis or forward problem is significantly easier than solving the synthesis or inverse problem. Another classification of engineering simulation models is based on whether the models are deterministic or stochastic. A key issue in using simulation models to design engineered systems revolves around making predictions about the behavior and performance of these systems. While deterministic simulations may be useful in recreating past events or testing a hypothesis in the presence of perfect knowledge, any simulation model used for prediction of the future needs to consider the uncertainty involved in system performance in order to be meaningful. Another classification of simulation models is based on the underlying mathematical and computational constructs used to model the engineered system. The constructs used to model engineered systems range from partial differential equations used to model continuous processes, such as material removal and bulk reactions, to discrete event models used to capture the event-driven nature of shop-floor and enterprise-level decisions. Finally, engineering simulations differ significantly based on how the results from the models are used by the decision-makers. Offline models are not as conservative about the use of computational resources and time requirements as online or real-time models, where latency in obtaining results is a critical factor and data from the real-world observations can alter the functioning of the models during execution (Douglas and Deshmukh, 2000).

In the past, the unavailability of computational resources to model most real-world engineered systems at a meaningful level of granularity has been a major constraint. The computational capabilities, in terms of compute cycles, data storage, and network bandwidth, offered by the next-generation cyberinfrastructure are quickly approaching the exascale range. This provides unprecedented opportunities to model engineered systems at a level of fidelity where meaningful predictions can be made in actionable time frames (Atkins *et al.*, 2003).

7.2 Engineering Simulation Highlights

The sites visited by the WTEC panel presented examples of the use of simulation and modeling in a variety of engineering domains and application areas (see the Asia and Europe Site Reports sections on the International Assessment of Research and Development in Simulation-Based Engineering and Science website, http://www.wtec.org/private/sbes/. The required password is available from WTEC upon request.) The examples highlighted below represent only a subset of the research conducted at the sites visited by the WTEC panel, which in turn represent only a subset of all modeling and simulation activities across engineering. These examples were selected to emphasize the diversity of application areas at the sites visited; the reader should not infer that these sites were conducting higher-quality research than those *not* included in this report. It is also important to note that given the scope of the study, most of the sites visited focused on simulation and modeling of physical processes and devices. Modeling of systems-level issues, involving logical and discrete event interactions, and the use of simulation output in decision-making, prediction, and design were not investigated in this WTEC study.

7.2.1 *Energy Systems*

The use of simulation and modeling related to energy systems revolves around several themes, such as combustion, wind turbines, fuel cells, hydrogen storage, photovoltaic interfaces, biofuel processes, nuclear reaction, and waste containment (ESF, 2007).

The Central Research Institute of Electric Power Industry (CRIEPI), a Japanese nonprofit corporation founded in 1951 with a broad mission of "solving global environmental problems while ensuring energy security," is conducting research on simulation analysis and tools for the Japanese electric power companies. Several examples of CRIEPI's activities are its Power system Analysis Tool (CPAT) for electric grid stability, computational fluid dynamics (CFD) analyses of combustion in coal-fired power plants, CFD analysis of transmission line vibration, and seismic analysis of oil tanks. For the most part, CRIEPI employs commercial software in its routine simulations, and it outsources code development to tailor the commercial tools for problems of interest.

Professor Nobuhide Kasagi in the Department of Mechanical Engineering at the University of Tokyo is heading a project on simulation of microturbines as mobile energy systems. His work focuses on developing a microturbine 30 kW energy generation system combined with a solid oxide fuel cell that would address the power supply needs for mobile systems.

Significant research is being conducted in Denmark on modeling and simulation of wind turbines. The Danish focus on this research area is quite understandable since the amount of wind power generated in Denmark per 1000 inhabitants exceeds 570 kW, which is by far the highest in Europe and, indeed, the world. Many advances in simulation-based design of wind turbines are associated with researchers at the Technical University of Denmark (DTU) and at Risø DTU National Laboratory for Sustainable Energy. The research at DTU is focused in the following areas: design of optimum airfoils, dynamic stall, especially 3-D stall, tip flows and yaw, heavily loaded motors, and interference (wake and park) effects. A few examples of successful application of basic research at DTU are the development of the popular aeroelastic code FLEX and the design of parts of the Nible and Tjaereborg turbines. In particular, the FLEX code has undergone developments over many generations (currently at FLEX5) and is used for designing wind turbines and analyzing loadings.

The Institut Français du Pétrole (IFP), a state-owned industrial and commercial establishment in France, is focusing on advancing research in energy, transportation, and the environment, and catalyzing the transfer of technology from fundamental research to industrial development. IFP's research activities in the energy systems area are targeted towards extended reserves, clean refining, and fuel-efficient vehicles. Extended reserves are based on the assumption that oil and other fossil fuels will remain the dominant source of transportation fuels and chemical feedstock; the research themes in this area target increasing the success rate in exploration, improving the recovery ratio in reservoirs, and developing new fields in extreme environments. Clean refining focuses on obtaining the highest possible yields of transport fuels from a unit basis of raw materials in an environmentally responsible fashion; the research themes in this area are the production of high-quality fuels; the conversion of heavy crudes, residues, and distillates; and the production of petrochemical intermediates. The fuel-efficient vehicles program recognizes the importance of reducing fuel consumption and the development of new powertrain systems for alternative fuels (e.g., biofuels), the research themes in this area are adaptive simulation of combustion and load balancing to develop highly efficient engine technologies, including conventional and hybrid powertrains, development of pollutant after-treatment technologies, development of electronic control strategy and onboard software, and validation and specification of alternative fuels (e.g., biofuels and NGV) with low CO_2 emissions.

Researchers in the United Kingdom are focused on developing a virtual power plant that will enable safe production of high-quality and cost-effective energy while extending the lifespan of power plants to 60 years, guaranteeing optimum fuel use, and better waste management. These challenges demand access to petascale machines to perform advanced simulations, along with development of a new generation of codes and simulation platforms (EPSRC, 2006).

7.2.2 *Disaster Planning*

Simulation and computational models are often used to evaluate civil structures and critical infrastructures in the event of disasters. These models are also needed to devise mitigation and rescue strategies. Examples of emergencies and critical infrastructures that are modeled include environmental phenomena such as earthquakes in a high-population-density urban environment; fires in large buildings and public facilities; chemical, biological, or radiation contamination from hazardous spills (or terrorism); large, near-shore oil spills in environmentally sensitive locations; and major floods and tidal waves (tsunamis). A key requirement in developing successful mitigation and rescue strategies is the ability to "predict" sufficiently accurately and in advance the specific events that may characterize the evolution of an emergency. The only way that this can be achieved is to run realistic simulations of the emergency in real time (EPSRC, 2006).

Significant research is being conducted at the University of Tokyo on modeling the impact of various natural and man-made disasters on large-scale infrastructures. Professor Shinobu Yoshimura's group is developing multiscale and multiphysics simulations for predicting quake-proof capability of nuclear power plants. The main objective of this project is full-scale simulation of components, but it also considers buildings subject to earthquakes. Most of the simulations involve one-way coupling on flow-structure interactions. A fully coupled simulation will require access to the next-generation supercomputer system at RIKEN. Professor Masaru Zako is leading a project on simulating disasters in chemical plants. The goal of this project is the development of a simulation capability for disaster propagation in chemical plants considering the effects of tank fire and wind speed and direction (see Fig. 7.1). Another key question addressed

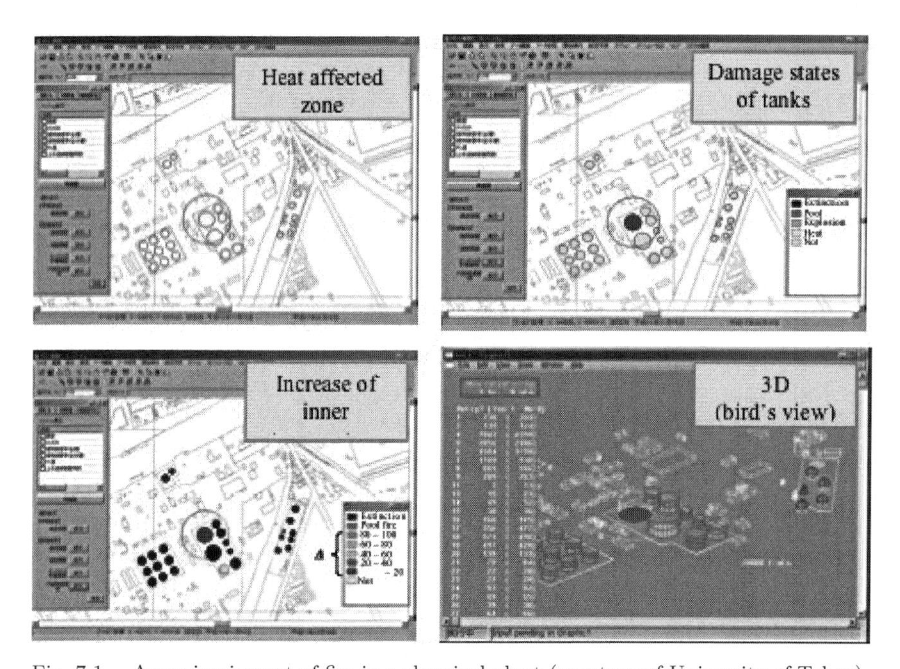

Fig. 7.1. Assessing impact of fire in a chemical plant (courtesy of University of Tokyo).

by this research is the estimation of associated damages using the results of simulations in conjunction with information from the Geographical Information System (GIS).

Professor Shunichi Koshimura's group at the University of Tokyo is working on developing fragility functions for tsunami damage estimation using numerical models and post-tsunami survey data from Banda Aceh, Indonesia. The purpose of this project is simulation of tsunamis and also estimation of damage caused based on simulation results and survey data. Advanced methods in solving the shallow water equations were developed for realistic terrains, and a "fragility function" that estimates the probability of damage was constructed for first time for Banda Aceh.

Earth Simulator Center (ESC) of the Japan Agency for Marine-Earth Science and Technology has been a leading research center on global climate simulation. The Earth Simulator Project has achieved resolution of 10 km, that can be further refined to 1 km resolution via adaptive mesh refinement. The ESC researchers validate the model with historical data. The group has also achieved some impressive results for weather prediction, primarily due to the speed at which predictions can be generated by the Earth Simulator. The group can obtain predictions for wind stream between buildings in downtown Tokyo for use in urban planning.

7.2.3 *Product and Process Modeling*

Development of new engineered product offers several opportunities for the use of simulation and computational modeling. A typical design and manufacturing cycle requires the product first to be modeled using computer-assisted design (CAD) software, either as a solid model or a B-rep model, which can be used for geometric analysis. The same product then needs to be represented in computer-assisted engineering (CAE) software, typically as a finite element mesh, which can be used to conduct distortion analysis during and after manufacturing. The model then is represented in computer-assisted manufacturing (CAM) software, typically as a solid model, to determine the optimal tool path for machining. Finally, the machine tool controller instructions need to be generated using the tool path model to drive a computer numerical control (CNC) machine to complete the manufacturing process. The current state of the art in different areas of design and manufacturing models also differs significantly. For example, finite element models of metal removal and forming processes (as shown in Fig. 7.2), which incorporate high strain rate, large deformation

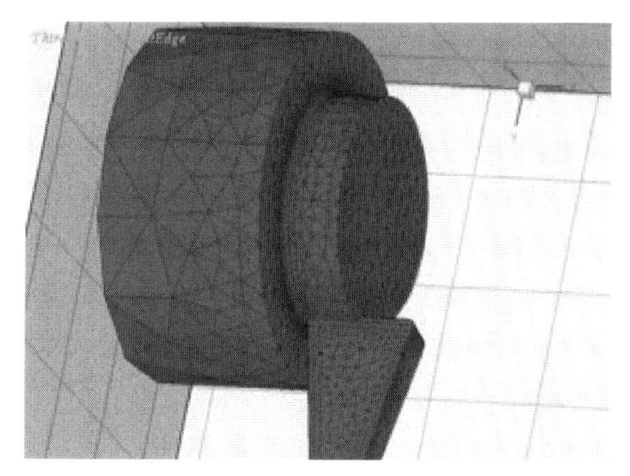

Fig. 7.2. FEA model of metal cutting/chip removal (courtesy of Third Wave Systems).

constitutive models, adaptive meshing, generalized contact algorithms, and heat transfer, can handle multiple length scales (10^{-4} m to 10^{2} m), and analysis of short (10^{-6} s) and long temporal (10^{3} s) events. In contrast, the modeling tools to support conceptual design are extremely primitive.

In a typical enterprise, the product development lifecycle consists of a sequence of complex tasks and decisions spread over a period of time. These tasks are either virtual, as in design, or physical, as in manufacturing and maintenance operations. Nevertheless, the success (both economic and operational) of an engineered system hinges on accurate prediction of the product performance during each stage of the lifecycle. For a successful product, all the decisions and operations have to be based on sound expectation of the implications of current decisions on downstream tasks. Most, if not all, of these tasks are performed in an *open-loop* fashion, i.e., very little feedback is available to any decision-maker on the implications. For example, consider the issue of level of permissible noise in a car. The level of noise tolerated in a small car in Europe is higher than in the United States. Initial designs of small cars were done for the European market. When the small car was introduced to the United States, the issue of isolation of noise was discovered to be a critical problem. Packaging or other simple redesign attempts were not able to resolve it. This problem could have been addressed in the initial design phase easily if it had been an explicit constraint in the optimal design formulation. However, the lack of information about the U.S. market and uncertainty about the importance of this issue led to expensive product redesign and manufacturing changes.

Toyota Central R&D Labs (TCRDL) has significant research activities focusing on issues related to automobile production, such as resource and energy conservation, environmental preservation, enhancement of comfort and safety, and advanced information processing. The highly publicized simulations of the structural responses of Toyota automobile frames to impacts, and simulations of the wind resistance levels of auto bodies, have reached a level of maturity such that they are no longer conducted at the TCRDL but at the Toyota Motor Company. Toyota has developed EVAS: Engine Vibration Analysis System, which is an engine design tool that is helping to develop lighter and quieter engines by providing highly accurate vibration predictions. The Aerodynamic Noise Simulator COSMOS-V developed at TCRDL is used to calculate temporally fluctuating airflow and to predict noise such as wind noise and wind throb, as well as vibrations caused by pressure fluctuations. Computational resources at Japan's Riken have also been used to conduct multiphysics simulations that combine airflow, heat distribution, structural deformations, and noise analysis for automobiles. Additionally, TCRDL has developed Mill-Plan software to determine the optimal number of machining processes and the tool form and cutting conditions for each process by closely examining the machining procedures and evaluating their overall efficiency.

In the safety/human engineering area, TCRDL is studying human body dynamics to determine the kind of injuries people are likely to suffer in auto collisions in order to help design safer automobiles. THUMS (Total HUman Model for Safety) is a suite of detailed finite element models of the human body that can be subjected to impacts to assess injury thresholds. Models included detailed meshing of musculoskelature and internal organs. It was evident to the WTEC panelists that these and other simulation activities are a result of TCRDL management's growing interest in modeling human behavior and interaction with machines, and social network interactions to understand disruptions to Toyota's production lines and/or to predict the behavior of markets towards disruptive technologies such as the hybrid automobile.

The European Union has several initiatives focused on developing complex engineered systems using computational models and simulation. For example, the European helicopter industry has a goal of developing a complete helicopter simulation that involves CFD-based simulations of aerodynamics, aeroacoustics, and dynamics of rotorcraft. The EC ACARE 2020 Green Aircraft directive provides the politically agreed-on targets for an acceptable maximum impact of air traffic on people and the environment, while at the same time allowing for a constantly increasing amount of air travel. The goals include a considerable reduction of exhaust gas and noise while air traffic increases by a factor of 3; accidents are expected to go down by 80%; passenger expense should drop by 50%; and flights must become largely weather-independent. Significant advances in multiscale and multiphysics modeling are needed to develop products that have a chance of achieving these goals.

Simulation and modeling also play a critical role in engineering one-of-a-kind devices for scientific exploration. CERN, The European Center for Nuclear Research, used simulation in the form of finite element modeling (FEM) at the beginning of construction of the Large Hadron Collider (LHC) to estimate the deflection expected in the detector itself. Using this approach, the maximum deflection (sag) in the 22-m-diameter detector, caused by it own weight of 7000 tons ($\sim$9 kN), was predicted to be 25 mm, which required compensation during the design stage. Accurate estimation of deflection to the cabin was critical to achieve the required alignment of the proton beam of the LHC. This example of civil engineering, in which the mass and volume of the detector was compensated by the removal of bedrock, is critical to the success of the forthcoming experiments. This was the first application of simulation to the engineering of the LHC.

Professor Gumbsch at the Fraunhofer Institute for Mechanics of Materials (Institut Werkstoffmechanik, IWM) has developed microscopic, physics-based models of materials performance and then inserted these subroutines into FEM codes such as LSDYNA, ABAQUS, and PEMCRASH. Specific applications of the IWM approach include modeling of materials (micromechanical models for deformation and failure damage analysis), simulation of manufacturing processes (pressing, sintering, forging, rolling, reshaping, welding, cutting), and simulation of components (prediction of behavior, upper limits, lifetime, virtual testing). The IWM also has a strong program in metronomy at small scales, developing models of failure/fracture, coupling data from nano-indentation with engineering- and physics-based models of same. The IWM has a substantial simulation effort in powder processing/sintering, and it has demonstrated the full process history from filling to densification and subsequent prediction of materials properties. Given its strong affiliation with the automobile industry, the IWM has sophisticated programs in modeling springback in stamped components, rolling, forming, and friction. It also has developed a detailed "concurrent," multiscale model of diamond-like-carbon thin-film deposition. The model integrates models from Schrödinger's equation up to classical models of stress–strain (FEM) and captures such properties as surface topography, internal structure, and adhesion.

7.2.4 *Computational Fluid Dynamics (CFD)*

CFD has been one of the primary engineering application domains for simulation using high-performance computing. The importance of this domain comes from the enormous range of practical applications of CFD, and from the existence of several different approaches to CFD itself. At one end, Direct Numerical Simulation (DNS) can be used to solve the governing Navier–Stokes equations using high-accuracy numerical methods. On the other end, Reynolds-Averaged Navier–Stokes (RANS) simulation is a computationally inexpensive method for complex problems, which removes small scales of turbulence by simple averaging.

In Europe there are several centers of excellence for CFD research. Large investments and effort have been put into CFD research in the European Union over the last 20–30 years. As an example, in the aerospace sector, the European Union has funded via the Framework Programmes many Specific Targeted Research Projects (STREP) related to CFD research. The European Union research community has also focused on collaboration within the community through a number of fora. First, ERCOFTAC (European Research Community on Flow, Turbulence and Combustion) is a well-established organization, funded through member subscriptions, in the area of fluid mechanics, turbulence, and combustion. The European Union research community has also focused on developing community codes that would be of value across several projects and reduce duplication of effort. Table 7.1 gives a list of some of the community codes developed by European Union researchers.

The project of Ecole Polytechnique Federale de Lausanne Chair of Modeling and Scientific Computing (EPFL-CMCS) "Multiphysics in America's Cup," is a tour-de-force in fluid flow modeling. This was a formal collaboration of EPFL researchers with the designers of Italy's entry into the America's Cup sailing race, the *Alinghi*, with substantial proprietary information being shared by the designers. Commercial codes were used, initially Fluent, then CFX, because of the perceived superiority of the included turbulence models. The implementation required a supercomputer. (In order to iterate the design with the developers, a turnaround time of 24 h was required.) The simulations had approximately

Table 7.1. European Community codes in engineering disciplines.

Name	Scientific area	Brief description licensing	# Users
Code_Aster	Engineering	Mechanical and thermal analysis free	300 (EDF) + 22000 Download since 2001
Code_Saturne	Fluid dynamics	Incompressible + free expandable fluids + heat transfer + combustion	80 (EDF) + 20 Industrial/ academic partners + 5 for teaching
OpenFOAM	Fluid dynamics + structural mechanics	Finite volume on free (fee for an unstructured support grid)	~ 2000
Salome	Framework for multiphysics simulations	Actually used for free engineering applications	50 (EDF) + 21 Institutions

20 million elements, with 162 million unknowns in the solution. The work was rewarded with two America's Cup victories.

Professor Leonhard Kleiser at the Computational Collaboratory at ETH Zurich is conducting research on turbulent flows and laminar-turbulent transition. Turbulent and transitional flows are investigated by direct numerical simulations. Here no turbulence models are employed, but the basic equations of fluid dynamics are numerically integrated on large computers whereby all relevant length and timescales of the flow must be resolved. Such investigations contribute to a better understanding of the fundamental phenomena and mechanisms of transition and turbulence, with respect to the conceptual exploration of methods for flow control and to the development of improved models for practical calculation methods.

The Institute of Fluid Mechanics of Toulouse (IMFT) conducts research in the mechanics of fluids with close experimental validation for applications in the areas of energy, processes, the environment, and health. The group of Dr. Braza, Director of Research of France's Centre National de la Recherche Scientifique (CNRS), has been working for a long time on direct numerical (DNS) and large-eddy simulations (LES) of turbulent prototype flows, and more recently on the detached eddy simulation (DES) of industrial-complexity flow applications. Specifically, DES was the central theme of a European initiative with the name DESider (Detached Eddy Simulation for Industrial Aerodynamics; http://dd.mace.manchester.ac.uk/desider), a (10 million program motivated by the increasing demand of the European aerospace industries to improve their CFD-aided tools for turbulent aerodynamic systems with massive flow separation. DES is a hybrid of LES and statistical turbulence modeling, and it is designed as a multiscale framework for modeling and capturing the inner and outer scales of turbulence. Furthermore, IMFT participates in the UFAST (Unsteady Effects in Shock-Wave-induced separation) project of the European Sixth Framework Programme, which is coordinated by the Polish Academy of Science (Gdansk) Institute of Fluid-Flow Machinery, and which includes about 18 partners. The objectives are to perform "closely coupled experiments and numerical investigations concerning unsteady shock wave boundary layer interaction (SWBLI)" to allow for feeding back numerical results to the experiments, and vice versa, for the sake of physics and modeling of compressibility effects in turbulent aerodynamic flows. Using RANS/URANS and hybrid RANS-LES methods, UFAST aims at assessing new methods for turbulence modeling, in particular for unsteady, shock dominated flow. UFAST investigates the range of applicability between RANS/URANS and LES for transonic and supersonic flows.

7.3 Summary of Key Findings

Simulation and modeling are integral to every engineering activity. While the use of simulation and modeling has been widespread in engineering, the WTEC study found several major hurdles still present in the effective use of these tools:

- Interoperability of software and data are major hurdles
  - Commercial vendors are defining de facto standards
  - Very little effort goes beyond syntactic compatibility
- Use of simulation software by nonsimulation experts is limited
  - Codes are too complicated to permit any user customization
  - Effective workflow methods need to be developed to aid in developing simulations for complex systems
- In most engineering applications, algorithms, software and data are primary bottlenecks
  - Computational resources (flops and bytes) were not limiting factors at most sites

  - Lifecycle of algorithms is in the 10–20-year range, whereas hardware lifecycle is in the 2–3-year range
- Visualization of simulation outputs remains a challenge
  - Use of HPC and high-bandwidth networks has exasperated the problem
- Experimental validation of models remains difficult and costly
  - Models are often constructed with insufficient data or physical measurements, leading to large uncertainty in the input parameters
  - The economics of parameter estimation and model refinement need to be considered
- Uncertainty is not being addressed adequately in many of the applications
  - Most engineering analyses are conducted under deterministic settings
  - Current modeling and simulation methods work well for existing products; however, they are not ideally suited for developing new products that are not derivatives of current ones
  - Existing models are mostly used to understand/explain experimental observations
- Links between physical- and system-level simulations are weak
  - There is very little evidence of atom-to-enterprise models
  - Enterprise-level modeling and decision-making are not coupled tightly with the process- and device-level models
- Engineers are not being trained adequately in academia to address simulation and modeling needs
  - Possession of a combination of domain, modeling, mathematical, computational, and decision-making skills is rare in program graduates

7.4 Comparison of U.S. and Worldwide Engineering Simulation Activities

- On average, U.S. academia and industry are ahead (marginally) of their European and Asian counterparts
  - Pockets of excellence exist in Europe and Asia that are more advanced than U.S. groups (Toyota, Airbus, University of Stuttgart)

- European and Asian researchers rely on the United States to develop the common middleware tools
  - Their focus is on application-specific software
- In the United States, the evolution from physical systems modeling to social-scale engineered systems lags behind that of Japan
  - Japan is modeling behavioral patterns of 6 billion people using the Life Simulator
- European universities are leading the world in developing curricula to train the next generation of engineering simulation experts (COSEPUP, 2007)

References

Atkins, D. E., Droegemeier, K. K., Feldman, S. I., Garcia-Molina, H., Klein, M. L., Messerschmitt, D. G., Messina, P., Ostriker, J. P. and Wright, M. H. (2003). *Revolutionizing Science and Engineering through Cyberinfrastructure: Report of the National Science Foundation Blue-Ribbon Advisory Panel on Cyberinfrastructure* (National Science Foundation, Arlington, VA), http://www.nsf.gov/od/oci/reports/toc.jsp.

Committee on Science, Engineering, and Public Policy (2007). *Rising above the Gathering Storm: Energizing and Employing America for a Brighter Economic Future* (National Academies Press, Washington, DC), http://www.nap.edu/catalog.php?record_id=11463#orgs.

Douglas, C. and Deshmukh, A. (2000). *Dynamic Data Driven Application Systems: Report of the NSF Workshop, March 2000* (National Science Foundation, Arlington, VA), http://www.nsf.gov/cise/cns/dddas/.

Engineering and Physical Sciences Research Council of the UK, High End Computing Strategy Committee, High End Computing Strategic Framework Working Group (2006). *A Strategic Framework for High End Computing*, http://www.epsrc.ac.uk/ResearchFunding/FacilitiesAndServices/HighPerformanceComputing/HPCStrategy.

——— (2006). *Challenges in High End Computing*, http://www. epsrc.ac.uk/ResearchFunding/FacilitiesAndServices/HighPerformanceComputing/HPC-Strategy.

European Science Foundation, European Computational Science Forum (2007). *The Forward Look Initiative. European Computational Science: The Lincei Initiative: From Computers to Scientific Excellence*, http://www.esf.org/activities/forward-looks/all-current-and-completed-forward-looks.html.

Oden, J. T., Belytschko, T., Hughes, T. J. R., Johnson, C., Keyes, D., Laub, A., Petzold, L., Srolovitz, D. and Yip, S. (2006). *Revolutionizing Engineering Science through Simulation: A Report of the National Science Foundation Blue Ribbon Panel on Simulation-Based Engineering Science* (National

Science Foundation, Arlington, VA), http://www.nsf.gov/pubs/reports/sbes_final_report.pdf.

President's Information Technology Advisory Committee (2005). *Computational Science: Ensuring America's Competitiveness* (Executive Office of the President, Committee on Technology, Washington, DC), http://www.nitrd.gov/pitac/reports/.

Chapter 8

VERIFICATION, VALIDATION, AND UNCERTAINTY QUANTIFICATION

George Em Karniadakis

"... *Because I had worked in the closest possible ways with physicists and engineers, I knew that our data can never be precise...*" –Norbert Wiener

8.1 Introduction

Uncertainty quantification (UQ) cuts across many disciplines, from weather modeling, to human dynamics, to reservoir modeling, systems biology, and materials modeling, to name just a few areas. The sources of uncertainty may be associated not only with initial and boundary conditions, material properties, equations of state, loads, reaction constants, geometry, and topology, but also with constitutive laws. In time-dependent systems, uncertainty increases with time, hence rendering simulation results based on deterministic models erroneous. In engineering systems, uncertainties are present at the component, subsystem, and complete system levels; therefore, they are coupled and are governed by disparate spatial and temporal scales or correlations.

Designing safe and robust engineering systems requires sensitivity UQ and sensitivity analysis at all levels. Despite its importance in all branches of Simulation-Based Engineering and Science (SBE&S), the field of UQ is not well developed. However, recently there has been an intense interest in *verification and validation* (V&V) of large-scale simulations and in modeling and quantifying uncertainty, as it is manifested by the many workshops and special volumes in scientific journals (Karniadakis and Glimm, 2006; Schuëller, 2005; Ghanem and Wojtkiewicz, 2003; Karniadakis, 2002) that have been organized to address these issues in the last few years. The U.S. Defense Modeling and Simulation Office (DMSO, http://www.dmso.gov) of the Department of Defense (DOD) has been the leader in developing fundamental concepts as well as terminology for V&V. In 1994, DMSO published definitions of V&V that have been adopted by other professional engineering communities such as the American Institute of Aeronautics and Astronautics (e.g., see AIAA, 1998).

The DMSO language is very specific: *verification* is the process of determining that a model implementation accurately represents the developer's conceptual description of the model and the solution to the model. Hence, by verification we ensure that the algorithms have been implemented correctly and that the numerical solution approaches the exact solution of the particular mathematical model — typically a partial differential equation (PDE). The exact solution is rarely known for real systems, so "fabricated" solutions for simpler systems are typically employed in the verification process. *Validation*, on the other hand, is the process of determining the degree to which a model is an accurate representation of the real world from the perspective of the intended uses of the model. Hence, validation determines how accurate are the results of a mathematical model when compared to the physical phenomenon simulated, so it involves comparison of simulation results with experimental data. In other words, verification asks "Are the equations solved correctly?" whereas validation asks "Are the right equations solved?" Or as stated in Roache (1998), "verification deals with mathematics; validation deals with physics."

This V&V framework is not new; Fig. 8.1 shows a schematic of the simulation cycle proposed by the Society for Computer Simulation in 1979 (Schlesinger *et al.*, 1979), where two models are identified: a *conceptual* and a *computerized* model. The former is the mathematical model, whereas the latter is what is called today the "code." It is the output of the computerized model that is validated as shown in the schematic. A third stage, named "qualification" in the schematic, determines the adequacy of the conceptual model in providing an acceptable level of agreement for the domain of intended application. Hence, qualification deals with the broader issues of the correct definition of the system, its interactions with the environment, and so forth.

Validation is not always feasible (e.g., in astronomy or in certain nanotechnology applications), and it is, in general, very costly because it requires data from many carefully conducted experiments. To this end, characterization of experimental inputs in detail is of great importance, but of equal importance are the *metrics* used in the comparison, e.g., the use of low-order moments or of probability density function distributions (PDFs).

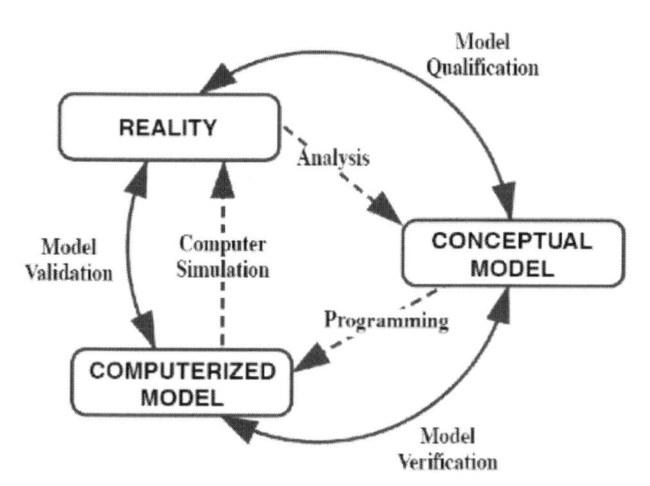

Fig. 8.1. Schematic illustrating the V&V stages as proposed in 1979 by the Society for Computer Simulation (Schlesinger *et al.*, 1979).

Validation involves many aspects from physics, experimentation, sensitivity analysis, and even philosophy of science. The comparison of simulation results with physical data represents to some extent a test of accuracy of a specific scientific theory. This interpretation is a topic of current debate, as there are many researchers who adopt the thinking of the twentieth-century science philosopher Karl Popper, who asserted that a scientific theory can only be invalidated but never completed validated! As Stephen Hawking further explains, "No matter how many times the results of experiments agree with some theory, you can never be sure that the next time the result will not contradict the theory" (Hawking, 1988).

It may be pragmatically useful to accept a well-tested theory as true until it is falsified, but this does not solve the philosophical problem of induction. Indeed, invalidation of a theory or a corresponding model can be done with a single point in the parameter space, but validation would require an infinite number of tests to cover the entire parameter space, an impossible task indeed! Nevertheless, experience has shown that in practical situations, comparisons of the simulations results with data from physical experiments are very useful and lead to the general acceptance of the model and its effective use in other similar situations. How accurate is the prediction of a model validated at one set of parameter values to another regime is an open scientific question, and no general guidelines exist. Hereafter, we will use the term "validation" in the conventional sense, and we direct the interested reader to the literature (Kleindorfer *et al.*, 1998) for other views on this subject.

Quantifying numerical inaccuracies in continuum or atomistic simulations is a difficult task; however, significant progress has been made, especially for PDE systems in *a posteriori* error estimation (Ainsworth and Oden, 2000). Uncertainty quantification in simulating physical systems is a much more complex subject; it includes the aforementioned *numerical* uncertainty, but often its main component is due to *physical* uncertainty. Numerical uncertainty includes in addition to spatial and temporal discretization errors, errors in solvers (e.g., incomplete iterations, loss of orthogonality), geometric discretization (e.g., linear segments), artificial boundary conditions (e.g., infinite domains), and others. Physical uncertainty includes errors due to imprecise or unknown material properties (e.g., viscosity, permeability, modulus of elasticity, etc.), boundary and initial conditions, random geometric roughness, equations of state, constitutive laws, statistical potentials, and others. Numerical uncertainty is very important and many scientific journals have established standard guidelines for how to document this type of uncertainty, especially in computational engineering (AIAA, 1998).

Physical uncertainty is much more difficult to quantify; it can be broadly characterized as *epistemic*, i.e., reducible, or as aleatory, i.e., irreducible. For example, given the current rapid advances in quantitative imaging technologies, we can expect that the rock permeability of an oil reservoir will be measured much more accurately in the future — this is an example of epistemic uncertainty. However, even in this case, and in many simulations of realistic configurations, uncertainty is irreducible beyond some level or scale, e.g., background turbulence — there are no absolutely quiet wind tunnels, and the atmosphere or the ocean are inherently noisy environments. For *atomistic* simulations, it should be appreciated that inter-atomic potentials even for simple media (e.g., water) are not known precisely.

Uncertainty modeling has been of interest to Operations Research for a long time, but there has been recently renewed interest in the role that uncertainty plays in *decision making*, where a distinction is made between situations where the probabilities are totally unknown as opposed to situations under risks with known probabilities. The former type of uncertainty is named Knightian uncertainty, in honor of Frank Knight who made this distinction in the 1920s (Starke and Berkemer, 2007). Another characterization is *deep uncertainty*, referring to the serious case where decision makers cannot even agree on the proper model or suitable cost function to be used (Lembert *et al.*, 2004). Moreover, uncertainty can be exogenous, because the behavior of other systems is uncertain, e.g., the effect of weather. This type of uncertainty quantification is of great interest not only to insurance and financial sectors but also to industry, including the automotive industry, as we discuss further in subsequent sections.

This chapter refers to "uncertainty" somewhat loosely in order to address diverse sources of stochasticity, e.g., from imprecise values of parameters in a physical system to inherent noise in biological systems or even to *variability* of parameters or operation scenarios. The following text gives a few examples of sources of uncertainty and subsequently reviews typical methods, the industry view, and finally summarizes the findings of the WTEC panel.

8.2 Effects of Uncertainty Propagation

A few examples from physical and life sciences illustrate the effects of uncertainty in the boundary conditions, material properties, and equation of state. In addition, inherent noise in the system, for example, in cellular and gene networks, may cause dramatic changes to the response.

8.2.1 *Fluid Mechanics*

The first example is on direct numerical simulations (DNS) of turbulence — a field established with the emergence of the first Cray supercomputers. Resolving the high wave number regime is typically a computationally challenging task, because it requires millions of grid points. This example, however, is from a comparison of DNS and experimental results in terms of the energy spectrum obtained in the wake of a circular cylinder (Ma *et al.*, 2000). There is agreement between the DNS and experiment in the high wave number regime, whereas there exists a rather large disagreement in the low wave number regime. This is surprising from the numerical discretization standpoint, but it is explained by the fact that for sufficiently high resolution (here more than 100 million degrees of freedom), the small scales are sufficiently resolved and do not depend on the boundary conditions, unlike the large scales which do depend on the boundary conditions. Since the inflow conditions in the experimental setup are not precisely known, simplified boundary conditions are employed in the DNS leading to this large disagreement.

8.2.2 *Plasma Dynamics*

The second example is from simulations of pulsed plasma microthrusters used in maneuvering satellites, and it involves ablation of Teflon®. The uncertainty here comes from the equation of state of Teflon® and its material properties. Results from two simulations using the same plasma code MACH2 (Kamhawi and Turchi, 1998) show two different equations of state. We observe that using the Los Alamos library SESAME in the MACH2 simulation, the pressure is almost 20 times greater than using another equation of state described in Kamhawi and Turchi (1998). The same magnitude in discrepancy has been demonstrated in another set of simulations (Gatsonis *et al.*, 2007) for the microthruster problem using Teflon® properties from different literature sources. The ablation here is modeled as a three-stage process that takes into account the phase transformation, two-phase behavior, and the depolymerization kinetics of Teflon®.

8.2.3 *Biomedical Applications*

In simulations of *biomechanics*, the values of the elastic properties of the vessel walls are important in order to capture accurately the blood flow — arterial wall interaction. The mean values of some of the required properties (e.g., modulus of elasticity, Poisson ratio) for the main arteries are tabulated in medical handbooks, but there is great variability among patients, even those with the same gender, age, or background.

In *systems biology*, uncertainty in initial conditions and reaction constants play a key role in the accuracy of the models, and hence sensitivity studies should be performed to assess this effect (Cassman *et al.*, 2007). Noise is inherent also at the gene level, and stochasticity has a defining role in gene expression because it may determine what type of phenotype is produced (Elowitz *et al.*, 2002; Pedraza and van Oudenaarden, 2005). Recent work has shown that even cells grown in the same environment with genetically identical populations can exhibit drastically different behaviors. An example is shown in Fig. 8.4 (right), from experiments with bacterial cells expressing two different fluorescent proteins (red and green) from identical promoters. Because of stochasticity (noise) in the process of gene expression, even two nearly identical genes often produce unequal amounts of protein. The resulting color variation shows how noise fundamentally limits the accuracy of gene regulation.

8.3 Methods

Most of the research effort in scientific computing so far has been in developing efficient algorithms for different applications, assuming an ideal input with precisely defined computational domains. Numerical accuracy and error control via adaptive discretization have been employed in simulations for some time now, but mostly based on both heuristics and in simple cases *a posteriori* error estimation procedures. With the field now reaching some degree of maturity, the interest has shifted to deriving more rigorous error estimators. It is also timely to pose the more general question of *how to model uncertain inputs and how to formulate new algorithms* in order for the simulation output to reflect accurately the propagation of uncertainty. To this end, the standard Monte Carlo (MC) approach can be employed, but it is computationally expensive and it is only used as the last resort. A faster version, the quasi Monte Carlo method (QMC), has been used successfully in computational finance and is particularly effective in applications with a relatively small *effective* dimensionality. The sensitivity method is an alternative, more economical, approach, based on moments of samples, but it is less robust and it depends strongly on the modeling assumptions. There are also other more suitable methods for physical and biological applications. The most popular technique for modeling stochastic engineering systems is the perturbation method, where all stochastic quantities are expanded around their mean value via a Taylor series. This approach, however, is limited to small perturbations and does not readily provide information on high-order statistics of the response. Another approach is based on expanding the inverse of the stochastic operator in a Neumann series, but this too is limited to small fluctuations. Bayesian statistical modeling has been used effectively in several different applications to deal with large uncertainties. At the heart of this framework is the celebrated Bayes' Theorem that provides a formal way of linking the prediction with the observed data.

A nonstatistical method, the polynomial chaos expansion (Ghanem and Spanos, 1991) and its extensions, has received considerable attention in the last few years, as it provides a high-order hierarchical representation of stochastic processes, similar to spectral expansions. Its main drawback is that currently it cannot handle high-dimensional problems; however, new algorithms under development aim to overcome this difficulty. One such approach is the use of sparse grids in multidimensional integration (Bungartz and Greibel, 2004) instead of full tensor-product forms, which in conjunction with direction-based adaptivity can result in very accurate and computationally efficient algorithms for problems with 100 parameters or so. The "curse of dimensionality" is a fundamental problem that currently has no effective treatment, and it is at the heart of the major computational bottleneck in stochastic modeling in many different applications. To this end, a unique new "priority program" was launched in 2008 by the German National Science Foundation (DFG) with focus on *high dimensions.* Specifically, the new priority program is entitled "Extraction of quantifiable information from complex systems," its duration is six years, and its aim is the development of new mathematical algorithms for modeling and simulation of complex problems related to high-dimensional parameter spaces; uncertainty quantification is one of the main thrusts in this program. Researchers in computational finance have also been making progress in dealing with many dimensions, and ideas from that field may also be applicable to physical and life science problems.

8.3.1 *Engineering Systems*

8.3.1.1 *Materials*

Another class of problems that illustrate the necessity of incorporating the effects of uncertainty (in the lower scales) is the design of components made of heterogeneous random media. Experimental evidence has shown that microstructural variability in such materials (e.g., polycrystalline materials) can have a significant impact on the variability of macroscale properties such as strength or stiffness as well as in the performance of the devices fabricated from such materials. It is well understood that the thermomechanical behavior at the device scale is significantly affected by variations in microstructure features. Texture variations have been found to significantly affect yield and ultimate tensile strength, bend ductility, and total fatigue life to failure. Similarly, grain-size variations have been found to affect ultimate and fracture strength of polycrystalline materials. No predictive modeling or robust design of devices made of heterogeneous media (MEMS, energetic materials, polycrystals, etc.) is practically feasible without accounting for microstructural uncertainty. The recent report by the National Academies of Science on Integrated Computational Materials Engineering (ICME) (NRC, 2008) stresses the importance of uncertainty quantification in material properties that may be obtained by either direct measurements or modeling, and also stresses the need for UQ in each stage of the design process (NRC, 2008).

8.3.1.2 *Uncertainty-Based Design*

One of the most important goals of incorporating uncertainty modeling in large-scale simulations is that it will lead to new *nonsterilized* simulations, where the input parameters and geometric domain have realistic representations. The simulation output will be denoted not by single points but by distributions that express the sensitivities of the system to the uncertainty in the inputs. This is a key element to reliability studies and will provide the first step towards establishing simulation-based certificates of fidelity of new designs. It will also be a valuable tool for experimentalists as it will quantify individual sensitivities to different parameters, thereby suggesting new experiments and instrumentation.

Uncertainty management for engineering *systems planning and design* is also of great significance (de Neufville *et al.*, 2004). In the more traditional approach, probabilistic design is employed to deal with the risk and reliability associated with a specific design. However, a new paradigm where one plans for uncertainty *explicitly* can endow the design with flexibility and robustness and lead to new architectures that are markedly different from designs created to simply meet fixed specifications, e.g., design codes or government regulations. This new paradigm will require investigation of scenarios where changes in economic and political events or shifts in customer preferences may take place in the future. Such a design paradigm change will require incorporation of cost and option analysis, so that system managers responsible for the operation of the final product are able to redirect the enterprise to avoid downside consequences or exploit upside opportunities. This new concept, termed *strategic engineering* by de Neufville *et al.* (2004), is illustrated in Fig. 8.2 using the Weibull distribution often employed in bridge engineering.

For example, robust design of a bridge would require examining loads around the mean value, whereas risk-based design would require structural integrity of the bridge at extreme loads as specified, e.g., by existing regulations (left tail of the distribution). In the new design paradigm, an extra step is taken to investigate the possible enhancement of the strength of the structure well beyond the existing regulations, in anticipation that a construction of a second deck of the bridge may be needed in a few years. The designer then has to perform a financial analysis to investigate if such enhancement is cost-effective at this juncture. Recent work by financial analysts has developed the theory for *real options*, i.e., options analysis applied to physical systems (Trigeorgis, 1996), and this holds the promise

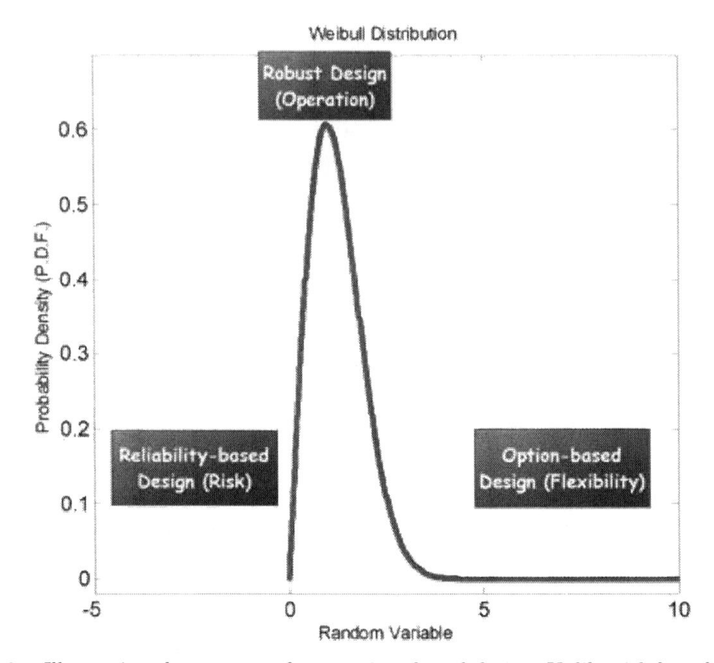

Fig. 8.2. Illustrating the concept of uncertainty-based design. Unlike risk-based design, where only the left side of the PDF is important, in the new paradigm the trade-off between flexibility and real options is studied by also examining the right tail of the PDF.

of enabling engineers to calculate the value of flexibility even at an early design stage. Uncertainty-based analysis is particularly important for the early stages of design in decision making, including acquisitions of expensive new products, e.g., for the Department of Defense.

8.3.1.3 *Certification/Accreditation*

For engineering systems, a *hierarchical* approach to validation is required to identify a set of possible experiments against which the simulation results from the various codes used will be tested. This is a complex and often expensive process, as it requires testing of the complete system. For example, in the aerospace industry two entire aircrafts should be tested for regulatory assessment. To be effective, the validation hierarchy should be *application-driven* rather than code-driven. It is common to construct a *validation pyramid* consisting of several tiers, as shown in Fig. 8.3. On the left hand of the validation pyramid are placed the selected experiments; on the right side are placed the simulation results to be compared against

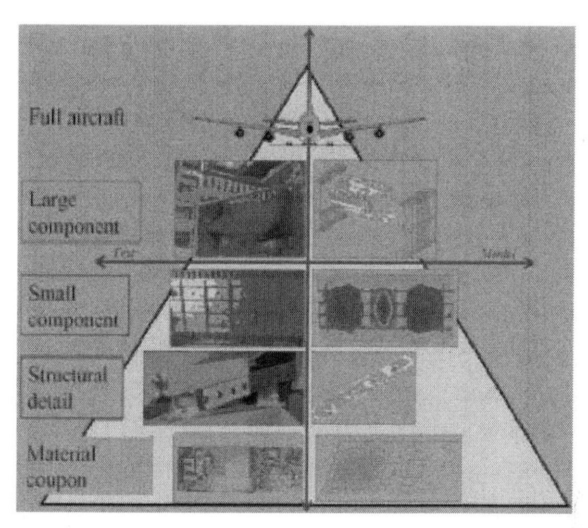

Fig. 8.3. Validation hierarchy for a full aircraft: on the left of the pyramid the selected experiments are placed; on the right are the corresponding simulation results. This is an idealized validation pyramid targeting aircraft structural design (Babuska *et al.*, 2007).

the experiments. Both the complexity and the cost of the experiments grow as we move up the pyramid. As we move down the pyramid, the emphasis changes from validating multiphysics codes to singlephysics codes. At the lowest tier, we need to identify simple-geometry experiments that can be tested by general codes, whereas as we move up the pyramid at the subsystem, system, and complete-system tiers, there exist multiscale and multiphysics coupling in complex-geometry domains, and it is much more difficult to select the required experiments. Each face of the pyramid corresponds to specific validation experiments for each computer code for each component of the design. The edges of the pyramid represent coupling between the different codes.

It is clear that the validation pyramid is not a unique construct and certainly not optimum for every application scenario of the complete system. However, it is in a sense an optimization problem, because the objective is to maximize reliability of the design given the constraint of a specific financial budget (Babuska *et al.*, 2007). The validation pyramid represents the engineering systems viewpoint in modeling and simulation-based design rather than the viewpoint of a specific discipline (Oberkampf *et al.*, 2004).

A case study of how the accreditation tests may lead to the rejection of the model is discussed by Babuska *et al.* (2007) for the Airbus A380 wing.

The wing test failed on February 14, 2006. The European Aviation Safety Agency (EASA) specifies that the wing has to endure a load that is 150% of the limit load for 3 seconds, but the wing failed at 147% of the limit load.

8.3.2 *Molecular Systems*

Verification and validation of atomistic simulations is not well developed. Most researchers perform routine checks of convergence of statistics, especially with respect to the size and number of time steps. This is particularly important in mesoscopic methods where large time steps are employed. For example, in density-functional theory (DFT) a further step in verification could involve benchmarking of DFT codes against standard community codes. UQ activities are rare, but some recent work at the Technical University of Denmark (DTU) (Mortensen *et al.*, 2005) has developed practical schemes based on Bayesian error estimation in DFT. This approach involves the creation of an ensemble of exchange-correlation functionals by comparing with an experimental database of binding energies of molecules and solids. The fluctuations within the ensemble are used to estimate errors relative to the experimental data for the simulation outputs in quantities such as binding energies, vibrational frequencies, etc. This procedure also checks for converged statistics as well as discretization and stationarity errors.

An example of the Bayesian error estimation in DFT is shown in Fig. 8.4. The two plots show binding energies versus bond lengths for some solids (left) and molecules (right). The green spots (the topmost spots in the legend) indicate experimental values while the crosses are calculated with a best-fit model from the class of gradient-corrected density-functional-theory (GGA-DFT) models. The "clouds" represent ensemble calculations predicting error bars on the calculated properties (shown as horizontal and vertical bars). The error bars are seen to provide reasonable estimates of the actual errors between best-fit model and experiment.

8.4 Industrial View

Industry, in general, is greatly interested in quantifying *parametric* uncertainty, and in major companies like BASF (Germany), ENI (Italy), and Mitsubishi Chemicals (Japan) that the WTEC panel visited (see the Asia and Europe Site Reports sections on the International Assessment of Research and Development in Simulation-Based Engineering and Science

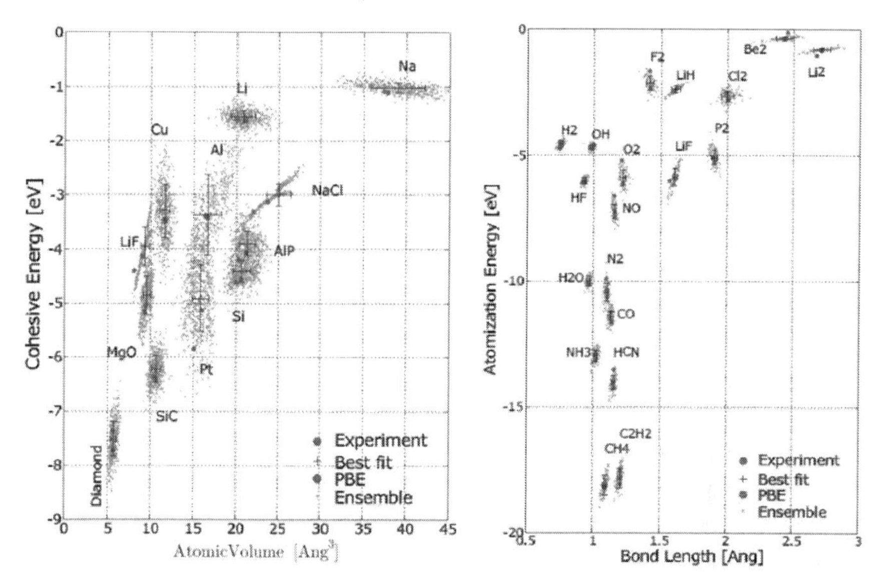

Fig. 8.4. GGA-DFT ensemble results for solids and molecules with uncertainty quantification (Mortensen *et al.*, 2005).

website, http://www.wtec.org/private/sbes/. The required password is available from WTEC upon request), this seems to be the main UQ-oriented concern. Researchers often design several experiments in order to obtain the range of the parameters in their models so they can perform their simulations. In electrical power systems, both in land projects or in new projects such as the all-electric ship (see websites of the Electric Ship Research and Development Consortium: http://www.esrdc.com and http://www.esdrc.mit.edu) or future Boeing airplanes, uncertainty modeling is of paramount importance, for estimating robustness both of the electric components and of the entire system and for evaluating risk and assessing various reconfiguration scenarios. A similar type of modeling is required in the auto industry as researchers there attempt to simulate the noisy RLC circuits in fuel cells. Uncertainties in such problems are due to load models, fault time, frequency, angle and voltage variations, operating conditions, and manufacturing variability of the different components.

An example from the chemical industry of the effect of the uncertainty in the thermodynamic properties on the design process is discussed by Larsen (1986). There are typically three stages in process design. In the first stage, where process screening takes place, only modest accuracy of properties is required. Physical properties are frequently estimated by engineering correlations or measured by simple, low-cost experiments (e.g., infinite dilution activity coefficients by gas chromatography). An accuracy of ±25% is acceptable in cost estimates. The demands for data accuracy vary: for example, a 20% error in density may result in a 16% error in equipment size/cost, whereas a 20% error in diffusivity may result in 4% error in equipment size/cost. The errors in density are usually small for liquids, but the errors in diffusivity are frequently large (factor of two or more). On the other hand, a 10% error in activity coefficient results in negligible error in equipment size/cost for easily separated mixtures, but for close-boiling mixtures (relative volatility <1.1) *a 10% error can result in equipment sizes off by factor of two or more.*

In Stages II and III of the design process, more precise and additional thermophysical properties data are required, so selective experiments may be commissioned at Stages II and III. In particular, Stage III design is typically performed using interpolated data based on experimental measurements. The cost estimates at the conclusion of Stage III is ±6% in this particular study (or ±3% if at engineering drawing stage). This case study is typical of the effect of parametric uncertainty in the design process, and it has a direct effect on the decision making process.

In addition to parametric uncertainties, other types of uncertainties are of interest to industry. For example, in the Toyota Central R&D Laboratory (CRDL), researchers are interested in quantifying uncertainty primarily in the context of human dynamics and decision making. To this end, they organized a workshop entitled "Prospects and Limitations of Mathematical Methods for Decision Making in Nonlinear Complex Systems" that took place in Denmark (Helsingor) 21–22 November 2006 (Starke and Berkemer, 2007). A second workshop on "Complex Systems: Interaction and Emergence of Autonomous Agents" was held in Baden, Austria in 2007. These workshops promoted the development and application of new methods for decision making and strategic planning, in particular with respect to uncertainty and nonlinear effects in complex systems through the discussion of a limited number of invited scientists with different research backgrounds from various fields. Their aim is to quantify uncertainty in both human activities and environmental issues.

The aerospace industry employs sensitivity analysis and hierarchical validation, such as validation pyramids, to quantify uncertainties both at the component and system level. In Europe, the MUNA (Management and Minimization of Uncertainties in Numerical Aerodynamics) initiative aims to quantify uncertainty in flow-structure interactions, e.g., in large aeroelastic motions, for aircraft technology. It is managed by the DLR Institute of Aerodynamics and Flow Technology in Braunschweig, with participation of eight university institutes, Airbus Germany, EADS-MAS, and Eurocopter. In four steps, error sources will first be identified, their effects will be quantified and made accessible to users via uncertainty margins, and finally strategies will be formulated to minimize simulation errors.

8.5 Summary of Findings

The NSF SBES report (Oden *et al.*, 2006) stresses the need for new developments in V&V and UQ in order to increase the reliability and utility of the simulation methods at a profound level in the future. A report on European computational science (ESF, 2007) concludes that "without validation, computational data are not credible, and hence, are useless." The aforementioned National Research Council report (2008) on integrated computational materials engineering (ICME) states that, "Sensitivity studies, understanding of real world uncertainties and experimental validation are key to gaining acceptance for and value from ICME tools that are less than 100 percent accurate." A clear recommendation was reached by a recent study on Applied Mathematics by the U.S. Department of Energy (Brown, 2008) to "significantly advance the theory and tools for quantifying the effects of uncertainty and numerical simulation error on predictions using complex models and when fitting complex models to observations."

The data and other information the WTEC panel collected in its year-long study suggests that there are a lot of "simulation-meets-experiment" types of projects but no systematic effort to establish the rigor and the requirements on UQ and V&V that the cited reports have suggested are needed. Overall, the panel found that the United States leads the research efforts at this juncture — certainly in terms of volume — in quantifying uncertainty, mostly in computational mechanics; however, there are similar recent initiatives in Europe. Specifically, fundamental work in developing the proper stochastic mathematics for this field, e.g., in addressing the high-dimensionality issue, is currently taking place in Germany, Switzerland, and Australia.

In the United States, the DOD Defense Modeling and Simulation Office has been the leader in developing V&V frameworks, and more recently the Department of Energy has targeted UQ through the ASCI program and its extensions (including its current incarnation, PSAAP). The ASCI/PSAAP program is focused on computational physics and mechanics problems, whereas DMSO has historically focused on high-level systems engineering, e.g., warfare modeling and simulation-based acquisition. One of the most active groups in V&V and UQ is the Uncertainty Estimation Department at Sandia National Labs, which focuses mostly on quantifying uncertainties in complex systems. However, most of the mathematical developments are taking place in universities by a relatively small number of individual researchers.

There are currently no funded U.S. national initiatives for fostering collaboration between researchers who work on new mathematical algorithms for V&V/UQ frameworks and design guidelines for stochastic systems. In contrast, there are several European initiatives within the Framework Programmes to coordinate research on new algorithms for diverse applications of computational science, from nanotechnology to the aerospace industry. In Germany, in particular, UQ-related activities are of utmost importance at the Centers of Excellence; for example, at the University of Stuttgart there is a Chair Professorship on UQ (sponsored by local industry), and similar initiatives are taking place in the Technical University of Munich and elsewhere.

On the education front, the WTEC panel found that existing worldwide graduate-level curricula do not offer the opportunity for rigorous training in stochastic modeling and simulation in any systematic way. Indeed, very few universities offer regular courses in stochastic partial differential equations (SPDEs), and very few textbooks exist on numerical solution of SPDEs. The typical graduate coursework of an engineer does not include advanced courses in probability or statistics, and important topics such as design or mechanics, for example, are taught with a purely deterministic focus. Statistical mechanics is typically taught in Physics or Chemistry departments that may not fit the background or serve the requirements of students in Engineering. Courses in mechanics and other disciplines that emphasize a statistical description *at all scales* (and not just at the small scales) due to both intrinsic and extrinsic stochasticity would be particularly effective in setting a solid foundation for educating a new cadre of simulation scientists.

References

American Institute of Aeronautics and Astronautics (1998). *AIAA Guide for Verification and Validation of Computational Fluid Dynamics Simulations* (AIAA, Reston, VA).

Ainsworth, M. and Oden, J. T. (2000). *A Posteriori Error Estimation in Finite Element Analysis* (John Wiley, New York).

Babuska, I., Nobile, F. and Tempone, R. (2007). Reliability of computational science, *Num. Methods for PDEs*, 23(4), 753–784.

Brown, D. L. (2008). *Applied Mathematics at the U.S. Department of Energy: Past, Present, and a View to the Future* (DoE, Washington, DC).

Bungartz, H.-J. and Greibel, M. (2004). Sparse grids, *Acta Num.*, 13, 10123.

Cassman, M., Arkin, A., Doyle, F., Katagire, F., Lauffenburger, D. and Stokes, C. (2007). *Systems Biology: International Research and Development* (Springer, Dordrecht, The Netherlands).

de Neufville, R., *et al.* (2004). Uncertainty management for engineering systems planning and design, *1st Eng. Sys. Symp.*, MIT.

Elowitz, M. B., Levine, A. J., Siggia, E. D. and Swain, P. S. (2002). Stochastic gene expression in a single cell, *Science*, 297, 1183–1196.

Gatsonis, N. A., Juric, D., Stachmann, D. P. and Byrne, L. (2007). Numerical analysis of Teflon ablation in pulsed plasma thrusters, *Mtg. Paper 43rd AIAA/ASME/SAI/ASEE Joint Prop. Conf. and Exhibit*, AIAA.

Ghanem, R. G. and Spanos, P. D. (1991). *Stochastic Finite Elements: A Spectral Approach* (Springer-Verlag, New York).

Ghanem, R. G. and Wojtkiewicz, S. F. (2003). Uncertainty quantification (special issue), *SIAM J. Sci. Computing*, 26(2).

Hawking, S. (1988). *A Brief History of Time* (Bantam Books, London).

Kamhawi, H. and Turchi, P. J. (1998). Design, operation, and investigation of an inductively-driven pulsed plasma thruster, *34th AIAA/ASME/SAE/ASEE Joint Prop, Conf. and Exhibit*, AIAA.

Karniadakis, G. E. (2002). Quantifying uncertainty in CFD, *J. Fluids Eng.*, 124(1), 2–3.

Karniadakis, G. E. and Glimm, J. (2006). Uncertainty quantification in simulation science, *J. Comp. Phys.*, 217(1), 1–4.

Kleindorfer, G. B., O'Neil, L. and Ganeshan, R. (1998). Validation in simulation; various positions in the philosophy of science, *Management Sci.*, 44, 1087–1099.

Larsen, A. H. (1986). Data quality for process design, *Fluid Phase Equilibria*, 29, 47–58.

Lembert, R., Nakicenovic, N., Sarewitz, D. and Schlesinger, M. (2004). Characterizing climate-change uncertainties for decision-makers: an editorial essay, *Clim. Change*, 65(2), 1–9.

European Computational Science Forum of the European Science Foundation (2007). *The Forward Look Initiative. European Computational Science: The Lincei Initiative: From Computers to Scientific Excellence.* http://www.esf.org/activities/forward-looks/all-current-and-completed-forward-looks.html.

Ma, X., Karamanos, G.-S. and Karniadakis, G. E. (2000). Dynamics and low-dimensionality of a turbulent near wake, *J. Fluid Mech.*, 410, 29–65.

Mortensen, J. J., Kaasbjerg, K., Frederiksen, S. L., Nørskov, J. K., Sethna, J. P. and Jacobsen, K. W. (2005). Bayesian error estimation in density-functional theory, *Phys. Rev. Lett.*, 95, 216401.

National Research Council of the National Academies (2008). *Integrated Computational Materials Engineering: A Transformational Discipline for Improved Competitiveness and National Security* (National Academies Press, Washington, DC).

Oberkampf, W. L., Trucano, T. G. and Hirsch, C. (2004). Verification, validation, and predictive capability in computational engineering and physics, *Appl. Mech. Rev.*, 57(5), 345–384.

Oden, J. T., Belytschko, T., Hughes, T. J. R., Johnson, C., Keyes, D., Laub, A., Petzold, L., Srolovitz, D. and Yip, S. (2006). *Revolutionizing Engineering Science through Simulation: A Report of the National Science Foundation Blue Ribbon Panel on Simulation-Based Engineering Science*, (National Science Foundation, Arlington, VA), http://www.nsf.gov/pubs/reports/sbes_final_report.pdf.

Pedraza, J. M. and van Oudenaarden, A. (2005). Noise propagation in gene networks, *Science*, 307, 1965–1969.

Roache, P. J. (1998). *Verification and Validation in Computational Science and Engineering* (Hermosa Publishers, Albuquerque, NM).

Schlesinger, S., Crosbie, R. E., Gagne, R. E., Innis, G. S., Lalwani, C. S. and Loch, J. (1979). Terminology for model credibility, *Simulation*, 32(3), 103–104.

Schuëller, G. I. (ed.) (2005). Special issue on computational methods in stochastic mechanics and reliability analysis, *Comput. Methods Appl. Mech. Eng.*, 194(12–16).

Starke, J. and Berkemer, R. (2007). Prospects and limitations of mathematical methods for decision making in nonlinear complex systems: synopsis of the workshop, *Decision Making and Uncertainty in Nonlinear Complex Systems*, Technical University of Denmark Department of Mathematics.

Trigeorgis, L. (1996). *Real Options: Managerial Flexibility and Strategy in Resource Allocation* (MIT Press, Cambridge).

Chapter 9

MULTISCALE SIMULATION

Peter T. Cummings

9.1 Introduction

The range of temporal and spatial time scales covered by simulation-based engineering and science (SBE&S), as illustrated in Fig. 9.1, is truly extraordinary. Even within a single subdomain, such as materials or biological systems modeling, the spatiotemporal range can be many orders of magnitude. In many applications, phenomena at one particular spatiotemporal scale have significance and impact at spatiotemporal scales both above and below. Hence, a major focus of SBE&S research, in all of its subfields, is the development of *multiscale simulation* methodologies — techniques that allow modeling of phenomena across disparate time and length scales. The increasing demands of engineering design to push beyond the limits of past engineering and scientific experience elevate the practical importance of achieving true multiscale simulation.

The term "multiscale simulation" is often loosely used to describe a wide variety of simulation activities in science and engineering. Broadly defined, all simulation could be classified as multiscale, since by definition simulation involves spanning at least one temporal and/or spatial scale. This is because in simulation we take a description of a phenomenon at one scale, derive the equations for the elements that interact at that scale and by solving dynamical equations of motion, predict the collective behavior of these elements as a system. For example, if we define models for how water molecules interact with each other, and solve the dynamical equations of motion for a large group of water molecules at the appropriate density and temperature, we can predict the properties of bulk water. In a second example, if we wish to understand the mechanical behavior of an engineering structure, such as an aircraft wing, we break it down into small components (finite elements) over which the mechanical properties are constant and behave in relation to each other in a well-defined way, and by solving the

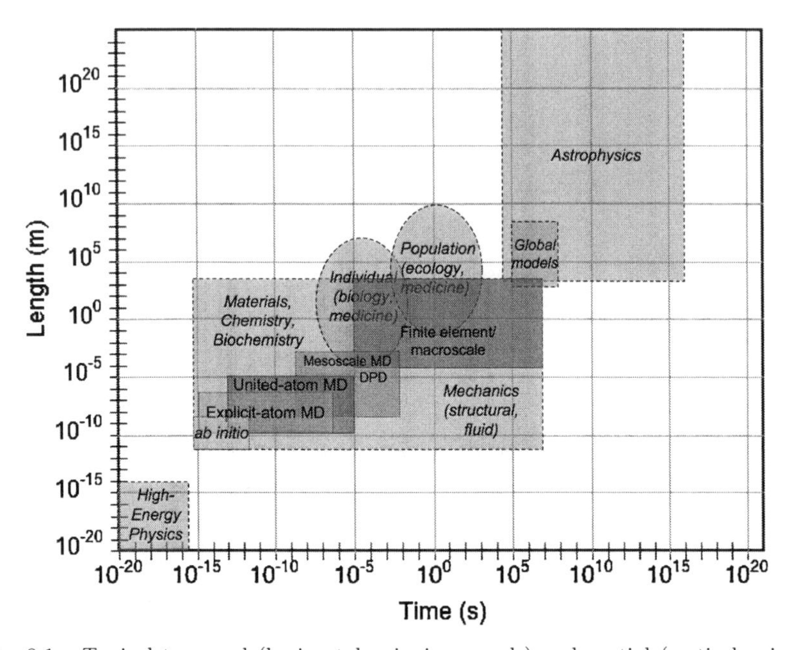

Fig. 9.1. Typical temporal (horizontal axis, in seconds) and spatial (vertical axis, in meters) scales for several SBE&S problem domains. High-energy physics is at the low end, with time scales set by subatomic particle collision and decay times, and spatial domains determined by subatomic particle size and relative distance. Material modeling methods, ranging from *ab initio* methods based on quantum mechanics, to explicit-atom molecular dynamics (MD, in which each atom is represented explicitly and its interactions and dynamics solved for), to united-atom MD (in which atoms clustered in functional groups, such as CH_2 and CH_3, are treated as single centers of force), to mesoscale MD and dissipative particle dynamics (in which whole molecules or clusters of molecules are frequently treated as a single entity, and, in the case of suspensions, the solvent degrees of freedom are integrated over to eliminate explicit solvent), and finally to finite element methods used to solve macroscopic balance equations, including the equations for structural mechanics. Also shown are the scales for modeling the biology of individual organisms and of populations, as well as astrophysics, where the low end is determined by the time scale (10^4 s) for a star spiraling into a million solar mass black hole with an innermost stable circular orbit of 10^8 m, and the upper end by simulations the age of the universe (6 billion years, or 10^{16} s). The time and spatial scales for global climate modeling of galaxy formation (a billion light years, or 10^{25} m, in width) over time scales of (labeled global models) are included for comparison.

dynamical equations that specify continuity of the appropriate properties within and at boundaries between the finite elements, we can simulate the mechanical response of the whole structure. In a third example, if we wish to predict how a cancerous tumor will grow and spread, one approach is to model it as a collection of cells, define the equations that govern the motion and state of each cell (single cell motion, cell–cell interactions, cell–environment interactions, cell growth and death), and by solving the dynamical equations for the cells, we can predict the behavior of the tumor. These are examples of simulation in the physical, engineering, and biological sciences, respectively.

More specifically, the foregoing are examples of *upscaling*, in which knowledge at a one level of description is obtained by averaging over dynamics at a level of description that is lower (i.e., more fundamental). Upscaling through one level of description is the workhorse of scientific and engineering simulation. However, we typically reserve the term "multiscale" for those simulations that involve multiple levels of upscaling, and that feature some degree of *downscaling* as well. By downscaling we mean that actions (or changes in boundary conditions) at one level of description of interest propagate down to the more fundamental levels.

As an example of downscaling, consider the simulation of an aircraft wing using finite elements (one level down). At sufficient levels of stress and fatigue, the possibility arises of a crack forming in part of the structure, near which point traditional macroscopic representations of the stress–strain relationship will fail. In order to model this at a high level of physical rigor, the finite elements subject to the highest strains may require modeling at a mesoscale level in which grain boundaries are taken into account. Finally, in the vicinity of the point at which a crack initiates, an explicit atomistic description is desirable, with the atoms at the point at which the crack initiates and propagates modeled via a first-principles (*ab initio*) method to accurately reflect the bond-breaking processes taking place at the point of rupture. Such a simulation, which involves many scales of description (*ab initio*, atomistic, mesoscale, and finite-element), in which information flows both upwards (upscaling) and downwards (downscaling), is an example of true multiscale materials simulation. Accordingly, an enduring goal in the materials modeling community is true multiscale simulation, in which the most fundamental levels of describing materials (electrons and nuclei, described by Schrödinger's equation) are connected seamlessly with the macroscopic scale to describe physical and chemical phenomena and devices at whatever level of detail is required for data-free prediction at some prescribed accuracy.

In biology, the range of scales is potentially just as great as or greater than the range of scales for materials. However, organizing principles at various levels (protein structure at the level of molecules, intracellular signaling at the level of single cells, genes, etc.) often mean that the most fundamental level in multiscale modeling of biological systems is at a higher level than the electronic level. Indeed, within biology, the term "*ab initio* protein folding" does not involve electronic structure calculations, as the name might imply to a materials simulator. *Ab initio* protein folding, also termed *de novo* protein folding, involves the prediction of protein structure from its molecular composition (amino acid sequence), as described in, for example, the reviews by Osguthorpe (2000) and Klepeis *et al.* (2002). The amino acid sequence is typically represented by an atomistically detailed model, or a coarse-grained version of this. The objective is to locate the structure with the global energy minimum. Thus, in principle, one approach to the problem of *ab initio* protein folding is molecular dynamics (MD) simulation. However, a fundamental problem in applying MD is the time scale for protein folding, which for even the simplest proteins is of the order of ms or longer. One of the heroic attempts to fold an atomistically detailed model for a protein in aqueous solution was that of Duan and Kollman (1998), who performed a 1-μs long MD simulation. In short, in biology, the most fundamental level is molecules (proteins, ligands, solvent, counter-ions), but many biological models begin with a cell as the fundamental building block (or, as in ecological modeling, a single organism). A hierarchy of the spatial and temporal scales relevant to modeling the onset and spread of cancer in humans is shown in Fig. 9.2.

Energy and environmental sustainability are areas in which multiscale simulation have played, and will continue to play, important roles. One notable and familiar example is global climate simulation (GCS), in which global circulation models (based on mass, energy, and momentum balances) for the oceans and the atmosphere are combined with models for relevant terrestrial processes in order to predict future climate trends. Two prominent examples of GCS codes are the United Kingdom's Meteorological Office Hadley Centre HadCM3 code, and the U.S. Geophysical Fluid Dynamics Laboratory's CM2 code. A large fraction of the computing cycles at several of the world's largest supercomputing center, including the Earth Simulator in Japan (http://www.jamstec.go.jp) and the National Center for Computational Sciences at Oak Ridge National Laboratory (http://nccs.gov), are devoted to running GCS codes, most often to investigate scientific questions but also to provide insight for policymakers. GCS may well be the application of SBE&S most familiar to the general public. The debate about the reliability of global warming simulation predictions has resulted in the general public becoming aware of predictive computational simulation, and that there can be potential economic

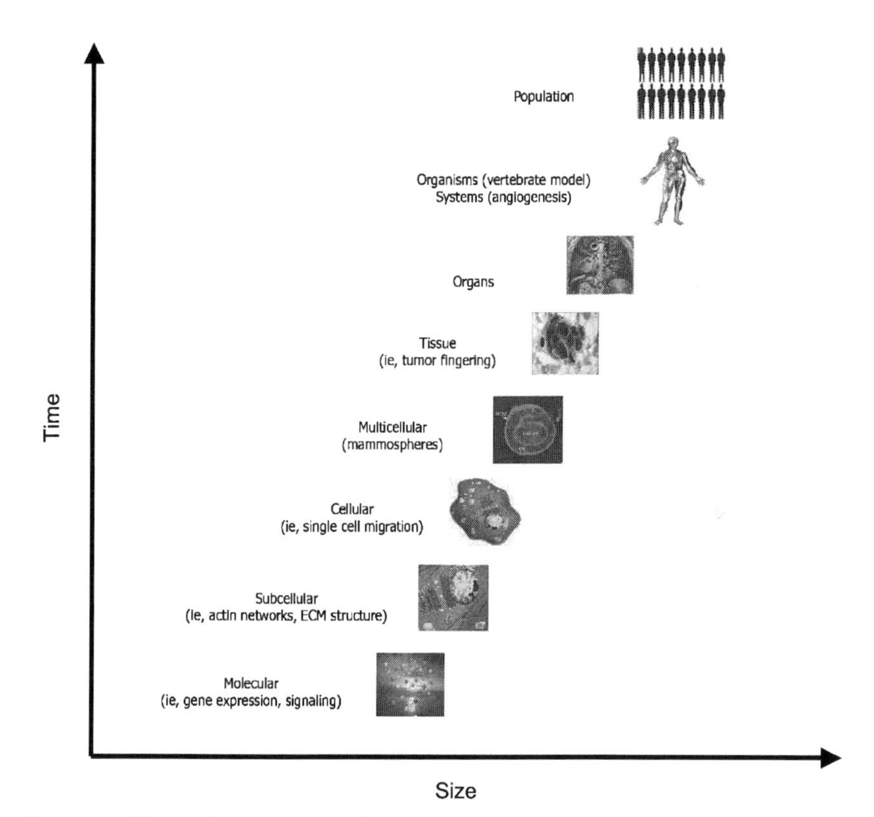

Fig. 9.2. A hierarchy of spatial and temporal scales relevant for modeling the initiation of cancer (oncogenesis) and its spread throughout the body (metastasis). At the highest level, bioinformatics is used to identify risk factors in populations, which can then be related through experiment to genetic abnormalities at the molecular level. The challenge in multiscale cancer modeling is to bridge between these two extremes to understand the mechanics of oncogenesis and metastasis, and identify effective patient-specific treatment strategies.

implications incurred by relying (or not relying) on the results of simulation. The predictions of global warming from GCS have large political and economic implications, and indeed the importance of GCS modeling is underlined by the 2007 Nobel Peace Prize being shared by the Intergovernmental Panel on Climate Change (http://www.ipcc.ch), much of whose work and conclusions are informed by GCS.

This chapter discusses some of the multiscale simulation issues and trends identified in the course of the WTEC international assessment. The following briefly reviews the state of the art, then presents examples of multiscale simulation viewed by the international assessment panel, and concludes with the findings of the panel in relation to multiscale simulation.

9.2 Current State of the Art in Multiscale Simulation

True multiscale simulations, with both automated upscaling and adaptive downscaling, are actually quite rare today. This is because there has been no broadly successful general multiscale modeling simulation methodology developed to date. Thus, multiscale simulation research efforts today could be said to fall into two main categories: methods designed to have general applicability in any multiscale simulation application, and multiscale simulation methods designed for a specific application or problem domain.

In specific problem domains, there are examples of successful applications of multiscale simulations. In crack propagation in metals, for example, multiscale simulation, spanning semiempirical *ab initio* methods, atomistic simulation, and finite element methods, was achieved in the last decade (Broughton *et al.*, 1999). Multiscale simulation methods, particularly relevant to nanoscale mechanical systems, are reviewed in a number of recent papers and monographs (Curtin and Miller, 2003; Liu *et al.*, 2004, 2006; Farrell *et al.*, 2007). The OCTA project (http://octa.jp) is another example of multiscale simulation modeling within a specific problem domain (polymer structure and dynamics); it is also a notable example of how a large government investment, combined with industry participation, can result in the development of a complex, integrated software environment useful both for fundamental research and for industrial application.

Among general multiscale simulation approaches, the equation-free modeling approach of Kevrekidis and coworkers (see, for example, Bold *et al.*, 2007; Kavousanakis *et al.*, 2007; Papavasiliou and Kevrekidis, 2007; Roberts and Kevrekidis, 2007) offers a framework for using existing modeling codes, enclosed with appropriate "wrappers," within a multiscale simulation framework. The equation-free modeling approach seems particularly suited to developing a coarse-grained model that computes properties as needed on the fly from simulations performed at a more detailed level. When implemented appropriately, it assumes no details of the model at the more detailed level, and uses standard integration of equations of motion at the coarse-grained level using properties computed at the more detailed level. Specific instances of this approach have been used before. For example, the Gibbs–Duhem method (Kofke, 1993) for computing phase equilibrium boundaries integrates a specific equation (the Gibbs–Duhem equation) to compute the saturated vapor pressure of a fluid by using two atomistic simulations (one of the vapor phase, one of the liquid phase) to determine the needed thermodynamic properties (differences in enthalpy and volume between the coexisting phases).

Despite advances in multiscale simulation, there is clearly much to be done in the future. The development of multiscale simulation methodologies, both general and specific, will continue to be the subject of intense effort, both in the United States and internationally. The national and industrial investment that led in Japan to Masao Doi's OCTA integrated simulation system for soft materials may provide insight into how domain-specific multiscale simulation methodologies can be realized.

9.3 Multiscale Simulation Highlights

This section highlights some of the multiscale simulations research being conducted at sites visited by the WTEC study panel. The reports below are not comprehensive, nor are their inclusion in this report meant to suggest that these sites conducted higher-quality research than other sites visited; rather, the intent is to choose a selection of sites that are representative of the range of activities observed during the WTEC study.

9.3.1 *Mitsubishi Chemical*

The Mitsubishi Chemical Group Science and Technology Research Center (MCRC), headquartered in Yokohama, Japan, is profiled in more detail in Chapter 3, Materials Simulation. It is also mentioned in this chapter because it exemplifies the industrial need for multiscale simulation, and because its efforts relate to specific findings.

MCRC staff use a variety of tools to design, construct, and operate optimal chemical and materials manufacturing processes. In order to achieve its design goals, MCRC couples commercial and in-house codes to create multiscale models of its processes, both existing and planned. Its researchers achieve this by writing their own in-house codes to provide coupling between the commercial codes. MCRC supports an internal multiscale molecular modeling effort on polymers, which is linked to the process modeling efforts. For MCRC staff, the lack of interoperability of commercial codes at different levels of spatio-temporal resolution makes the development of effective multiscale models difficult.

9.3.2 *Theory of Condensed Matter Group, Cavendish Laboratory, Cambridge University*

The Theory of Condensed Matter (TCM) group at Cambridge, headed by Mike Payne, is well known for the development of several DFT-based codes, notably the Cambridge Sequential Total Energy Package, (CASTEP, http://www.tcm.phy.cam.ac.uk/castep/) and the order-N electronic total energy package (ONETEP, http://www.onetep.soton.ac.uk). Additional information about these codes is available in the Europe Site Reports section of the International Assessment of Research and Development in Simulation-Based Engineering and Science website, http://www.wtec.org/private/sbes/ (required password available from WTEC upon request) and Chapter 3.

In addition to his efforts developing *ab initio* codes, Payne has been developing a multiscale simulation method for modeling of dynamics of materials, specifically crack propagation in graphine sheets, illustrated in Fig. 9.3. The technique used is called learn on the fly (LOTF) (Csányi *et al.*, 2004, 2005). As with many multiscale simulation methods that bridge between scales by constructing a hybrid Hamiltonian that combines different models, the LOTF method uses a unique short-ranged classical potential whose parameters are continuously tuned to reproduce the atomic trajectories at the prescribed level of accuracy throughout the system. Payne and coworkers have developed LOTF simulations with interfaces to a number of *ab initio* codes, in addition to the SIESTA code (Soler *et al.*, 2002) used to generate the configurations in Fig. 9.3.

9.3.3 *Blue Brain Project, École Polytechnique Fédérale de Lausanne*

The Blue Brain Project is a joint collaboration between the Brain Mind Institute at the École Polytechnique Fédérale de Lausanne (EPFL), Switzerland, and IBM, maker of the Blue Gene high-performance computing (HPC) platform. The success of the Blue Gene HPC platform is evident in the fact that four of the top 10 computers on the June 2008 top 500 supercomputer list (http://top500.org) were IBM Blue Gene architecture machines. In the same list, the IBM Blue Gene computer in the HPC center at EPFL (http://hpc.epfl.ch) was ranked 103.

The goal of the Blue Brain computational project, directed by Henry Markram and managed by Felix Schürmann and involving a team of 35 international scientists, is to model in great detail the cellular infrastructure

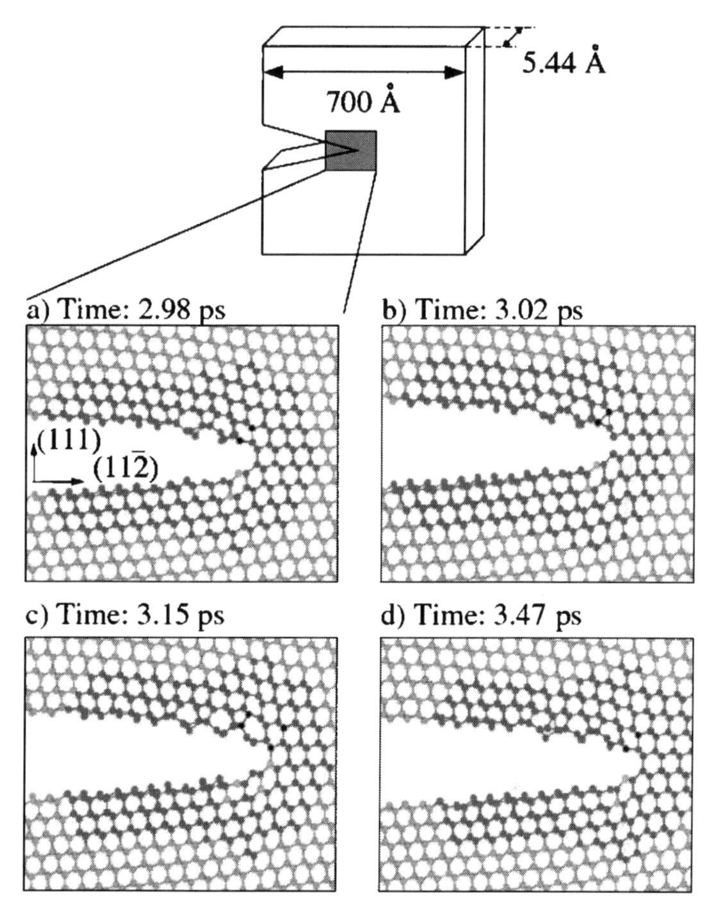

Fig. 9.3. Propagation of a crack in silicon using the LOTF scheme. Silicon atoms treated classically are shown in gold (gray), and those described using the SIESTA code are shown in red (black) (Csányi *et al.*, 2005).

and electrophysiological interactions within the cerebral neocortex, which constitutes about 80% of the brain and is believed to host cognitive functions such as language and conscious thought. Initiated by Henry Markram, an experimental biologist, the project started out with the goal of reverse-engineering the neocortex (Markram, 2006). Experimental data has been collected for different types of cells, electrical behaviors, and connectivities. Experimental imaging data is used to construct a faithful *in silico* replica at the cellular level of the neocortical column of a young rat. The detailed multiscale model consists of 10,000 neurons (involving approximately 400 compartments per neuron, using a Hodgkin–Huxley cable model), 340 types of neurons, 200 types of ion channels, and 30 million dynamic connections (synapses) (Kozloski *et al.*, 2008). The whole system is described by millions of ordinary differential equations modeling electrochemical synapses and ion channels. The main objective of the model is to explain the electrical, morphological, synaptic, and plasticity diversity observed in the laboratory, and to understand emergent network-level phenomena. Because of the nature of the model, it can be used both in an upscaling fashion (what network-level behavior emerges from a given cellular/connection structure?) as well as in a downscaling fashion (how do changes in the neocortical environment such as injury and drug regimens, and changes in physiological parameters such as electrolyte concentration, impact network behavior and individual neurons and synapses?).

Of additional note is another neocortex simulation project being pursued by the Computational Biology and Neurocomputing Group at the Royal Institute of Technology (CBNG-RIT) in Stockholm, Sweden (Djurfeldt *et al.*, 2008). At its largest incarnation, this simulation includes 22 million neurons and 11 billion synapses; it also is implemented on the IBM Blue Gene/L computing platform. The CBNG-RIT model involves a more coarse-grained description than the Blue Brain Project.

9.3.4 *Yoshimura Group, Department of Systems Innovation, School of Engineering, University of Tokyo*

The 21st Century Center of Excellence Program on Mechanical Systems Innovation in the School of Engineering of the University of Tokyo supports a number of projects that, if successful, will result in multiscale simulations relevant to energy and sustainability (see Asia Site Reports on the International Assessment of Research and Development in Simulation-Based Engineering and Science website, http://www.wtec.org/private/sbes/. The required password is available from WTEC upon request.). As one example, Shinobu Yoshimura, a faculty member in the Department of Systems Innovation, is developing a multiscale/multiphysics simulation for predicting the degree to which nuclear power plants can be made quake-proof. He is building on his expertise and prior experience in large-scale mechanical systems simulation (Yoshimura *et al.*, 2002; Kim *et al.*, 2004; Chang *et al.*, 2005; Yoshimura, 2006).

9.4 Summary of Key Findings

True multiscale simulation has long been a goal of SBE&S. It quite naturally arises as a desirable capability in virtually every application area of SBE&S. It particularly arises when the notion of design and/or control comes into play, since in this case the impact of changes in problem definition at the larger scales needs to be understood in terms of the impact on structures at the smaller scales. For example, in materials, the question might be how changes in the desirable performance characteristics of a lubricant translate into differences in molecular structure of the constituent molecules, so that new lubricants can be synthesized. In biology, the question might be what needs to happen at the cell–cell interaction level in order to control the spread of cancer. In energy and sustainability, the goal might be to understand how setting emissions targets for carbon dioxide at power plants in the United States may affect the growth and biodiversity of South American rain forests. Answering all of these questions requires a combination of downscaling and upscaling, and so fall in the realm of multiscale simulation.

The WTEC panel's key findings on multiscale simulation are summarized below:

- Multiscale modeling is exceptionally important. It holds the key to making SBE&S more broadly applicable in areas such as design, control, optimization, and abnormal situation modeling.
- Successful examples exist within narrow disciplinary boundaries; these include, but are not limited to
  - Crack propagation within materials
  - Computational neuroscience
  - Global climate modeling
- Attempts to develop general strategies have not yet succeeded. Some general mathematical principles exist for upscaling, such as homogenization theory (Cioranescu and Donato, 2000). As noted in the state-of-the-art section, the equation-free method of Kevrekidis and coworkers (Bold *et al.*, 2007; Kavousanakis *et al.*, 2007; Papavasiliou and Kevrekidis, 2007; Roberts and Kevrekidis, 2007) is another attempt to develop a general approach to multiscale modeling.
- Because of the immediate and unavoidable needs for design and control, industry performs multiscale modeling with varying degrees of success. The lack of standards-based interoperability of codes is major

impediment to them in linking together codes at various scales. This problem was cited by a number of the companies the WTEC panel visited. Interoperability of codes does not, in itself, result in a multiscale model. Code interoperability should be regarded as a necessary but not sufficient requirement for the development of general-purpose multiscale simulation capabilities.

- U.S. research in multiscale simulation is on a par with Japan and Europe. In fact, U.S. leadership in high-performance computing resources has helped the United States to be competitive in this area. This is because in order to validate a multiscale simulation methodology it is necessary to perform the simulation at full detail (i.e., using the most fundamental level of description) enough times to provide a validation data set. For example, to validate a crack propagation simulation, the whole problem should be simulated multiple times at the first-principles level to provide the set of data to which the multiscale simulation should be compared. However, one concern is that access to the largest computational facilities (such as the emerging petascale platforms) may be difficult for a project whose stated aim is to run a series of petaflop-level simulations in order to develop and validate a multiscale modeling methodology. Access to petaflop-level resources today is generally focused on solving a small set of problems, as opposed to developing the methodologies that will enable the solution of a broader range of problems.
- For the most part, U.S. research in multiscale modeling is diffuse, lacking focus and integration, and federal agencies have not traditionally supported the development of codes that can be distributed, supported, and successfully used by others. By contrast, the Japanese and European approach is to fund large interdisciplinary teams, such as the OCTA project in Japan and the Blue Brain project in Switzerland, often with the goal of distributing the product codes either in open-source or commercial form.
- Solving the multiscale simulation challenge is without doubt an interdisciplinary endeavor; however, the tradition of interdisciplinary collaboration leading to community solutions is much stronger in Europe and Japan than in the United States. Some of the issues related to this are discussed in more detail in Chapter 3.

References

Bold, K. A., Zou, Y., Kevrekidis, I. and Henson, M. (2007). An equation-free approach to analyzing heterogeneous cell population dynamics, *J. Math. Biol.*, 55(3), 331–352.

Broughton, J. Q., Abraham, F. F., Bernstein, N. and Kaxiras, E. (1999). Concurrent coupling of length scales: methodology and application, *Phys. Rev. B*, 60(4), 2391–2403.

Chang, Y. S., Ko, H. O., Choi, J.-B., Kim, Y.-J. and Yoshimura, S. (2005). Parallel process system and its application to steam generator structural analysis, *J. Mech. Sci. Technol.*, 19(11), 2007–2015.

Cioranescu, D. and Donato, P. (2000). *An Introduction to Homogenization* (Oxford University Press, Oxford).

Csányi, G., Albaret, T., Moras, G., Payne, M. C. and De Vita, A. (2005). Multiscale hybrid simulation methods for material systems, *J. Phys. Cond. Matter*, 17, R691–R703.

Csányi, G. (2004). "Learn on the fly": a hybrid classical and quantum-mechanical molecular dynamics simulation, *Phys. Rev. Lett.*, 93(17), art. no. 175503.

Curtin, W. A. and Miller, R. E. (2003). Atomistic/continuum coupling in computational materials science, *Model. Sim. Matls. Sci. Eng.*, 11(3), R33–R68.

Djurfeldt, M., Lundqvist, M., Johansson, C., Rehn, M., Ekeberg, Ö. and Lansner, A. (2008). Brain-scale simulation of the neocortex on the IBM Blue Gene/L supercomputer, *IBM J. Rsch. Dev.*, 52(1, 2), 30–41.

Duan, Y. and Kollman, P. A. (1998). Pathways to a protein folding intermediate observed in a 1-microsecond simulation in aqueous solution, *Science*, 282(5389), 740–744.

Farrell, D., Karpov, E. and Liu, W. K. (2007). Algorithms for bridging scale method parameters, *Comp. Mech.*, 40(6), 965–978.

Kavousanakis, M. E., Erban, R., Boudouvis, A. G., Gear, C. W. and Kevrekidis, I. G. (2007). Projective and coarse projective integration for problems with continuous symmetries, *J. Comp. Phys.*, 225(1), 382–407.

Kim, J. C., Choi, J. B., Kim, Y.-J., Choi, Y.-H., Park, Y.-W. and Yoshimura, S. (2004). Development of an integrity evaluation system on the basis of cooperative virtual reality environment for reactor pressure vessel 270–273, *Advances in Nondestructive Evaluation*, Pts. 1–3. (Trans Tech Publications Ltd., Zurich-Uetikon) pp. 2244–2249.

Klepeis, J. L., Schafroth, H. D., Westerberg, K. M. and Floudas, C. A. (2002). Deterministic global optimization and ab initio approaches for the structure prediction of polypeptides, dynamics of protein folding, and protein–protein interactions 120, *Computational Methods for Protein Folding* (John Wiley & Sons, Inc., New York) pp. 265–457.

Kofke, D. A. (1993). Direct evaluation of phase coexistence by molecular simulation via integration along the saturation line, *J. Chem. Phys.*, 98(5), 4149–4162.

Kozloski, J., Sfyrakis, K., Hill, S., Schuermann, F., Peck, C. and Markram, H. (2008). Identifying, tabulating, and analyzing contacts between branched neuron morphologies, *IBM J. Rsch. Dev.*, 52(1, 2), 43–55.

Liu, W. K., Karpov, E. G. and Park, H. S. (2006). *Nano Mechanics and Materials: Theory, Multiscale Analysis and Applications* (John Wiley and Sons, New York).

Liu, W. K., Karpov, E. G., Zhang, S. and Park, H. S. (2004). An introduction to computational nanomechanics and materials, *Comp. Methods App. Mech. Eng.*, 193(17–20), 1529–1578.

Markram, H. (2006). The Blue Brain Project, *Nat. Rev. Neuro.*, 7(2), 153–160.

Osguthorpe, D. J. (2000). Ab initio protein folding, *Curr. Opinion Struct. Bio.*, 10(2), 146–152.

Papavasiliou, A. and Kevrekidis, I. G. (2007). Variance reduction for the equation-free simulation of multiscale stochastic systems, *Multiscale Mod. Sim.*, 6(1), 70–89.

Roberts, A. J. and Kevrekidis, I. G. (2007). General tooth boundary conditions for equation free modeling, *Siam J. Sci. Comp.*, 29(4), 1495–1510.

Soler, J. M., Artacho, E., Gale, J. D., García, A., Junquera, J., Ordejón, P. and Sánchez-Portal, D. (2002). The SIESTA method for ab initio order-N materials simulation, *J. Phys. Cond. Matter*, 14(11), 2745–2779.

Yoshimura, S. (2006). Virtual demonstration tests of large-scale and complex artifacts using an open source parallel CAE system, ADVENTURE 110, *Advances in Safety and Structural Integrity 2005* (Trans Tech Publications, Ltd., Zurich-Uetikon), pp. 133–142.

Yoshimura, S., Shioya, R., Noguchi, H. and Miyamura, T. (2002). Advanced general-purpose computational mechanics system for large-scale analysis and design, *J. Comput. Appl. Math.*, 149(1), 279–296.

Chapter 10

BIG DATA, VISUALIZATION, AND DATA-DRIVEN SIMULATIONS

Sangtae Kim

10.1 Introduction

One hundred meters under the Franco–Swiss border, along a circular path 17 miles in circumference, the collision of opposing beams of protons streaming at 99.999999% of the speed of light generates 1 petabyte per second of raw data (see Fig. 10.1). The gold standard in data-intensive scientific cyberinfrastructure distills this massive stream of data to "merely" petabytes per year of interesting archival events for further scientific analysis in the quest for a fundamental understanding of the nature of our material universe. The smooth transition and acceptance of "petascale" in our scientific lexicon belies the staggering magnitudes — peta, or 10^{15}, is the scale that bridges the macroscopic and molecular worlds; consider that each byte in a 30-petabyte storage system would map one to one to each molecule in a 1 mm $\times$ 1 mm $\times$ 1 mm cubic air sample (note: 0.224 cc $= 6 \times 10^3$ peta and 0.00112 cc $= 3 \times 10$ peta). More than two thousand miles away, in the massive Khurais oil fields of Saudi Arabia, 100 injection wells ring an area of 2700 square miles (see Fig. 10.2). Sophisticated SBE&S reservoir models use over 2.8 million 3D images of the Khurais underground strata to dictate dynamically the real-time operation of injection and production wells (King, 2008). Faced with declining output from aging fields, the optimal development of Khurais is key to maintaining the "growth option" for output in the world's largest exporter of oil. These two examples, at opposite ends of the spectrum of fundamental research/applied technology, and the extremum between the quest for knowledge of the nature of the universe and the pragmatic concerns for energy to fuel global economic development, underscore the pivotal role of "big data" and data-driven SBE&S. Furthermore, consider that the flood of data in the post-genomic era is the daunting barrier to progress in life sciences research and the search

Fig. 10.1. Image of the Large Hadron Collider (source: http://lhc-milestones.web.cern.ch/LHC-Milestones/images/photos/ph07-1.jpg).

for cures. Perhaps more urgently, consider the tantalizing prospect of data-intensive SBE&S for fresh approaches to understanding the complex system dynamics in the global financial marketplace and creation of new and reliable tools for governmental financial policymakers. It is no wonder then that the popular science and technology media prescribe the data deluge of the petabyte era as the end of the scientific method (Anderson, 2008).

In our global travels for the SBE&S project and the search for leadership in big data issues, an overarching finding is that aside from notable exceptions as mentioned in the following pages, there is a significant gap in the level of commitment to data versus computational infrastructure. We use the term "infrastructure" broadly, to denote not just the hardware and software, but human infrastructure (training and education) as well. The gap is narrower in the life sciences disciplines and in industrial R&D laboratories, but these efforts are in turn limited by the overall gap across the SBE&S educational landscape. Therefore, this motivated the organizational scheme for this chapter on "big data:" we lead with the

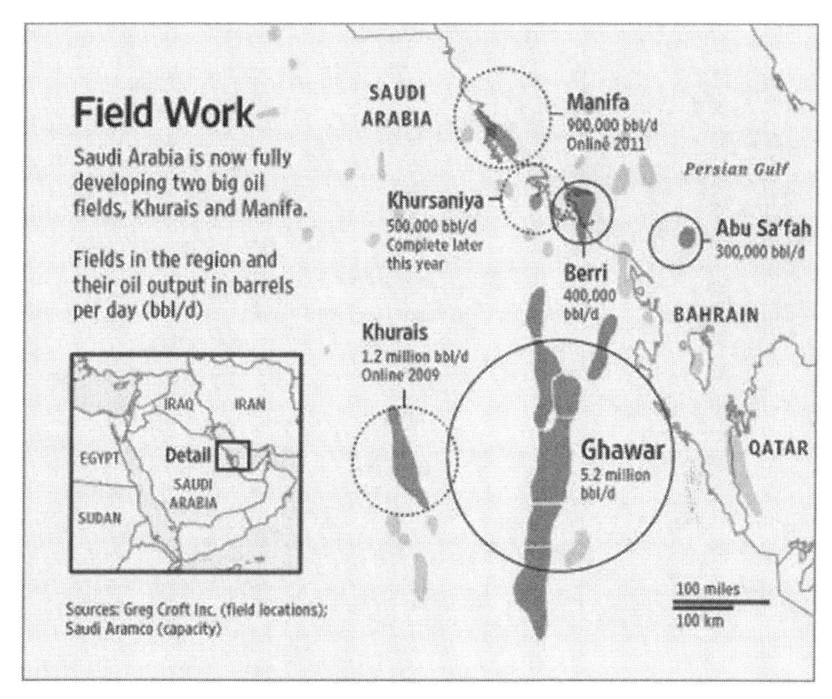

Fig. 10.2. Current and future oilfields in Saudi Arabia. (Source: *Wall Street Journal*, April 22, 2008.) The largest undeveloped oil fields including the Khurais (colored light gray) require significant SBE&S resources.

uniformly strong cyberinfrastructure organization of the particle physics research community, then describe the life sciences research examples across the globe, and conclude with an aggregate view of the enterprise scale data infrastructure and knowledge management/decision support systems in industrial R&D. We close the chapter with an overview of the (sorry) state of data infrastructure on university campuses and its impact on current capabilities for training and education.

10.2 Particle Physics Research: Petabytes Per Second

The particle physics research community and their search for the elusive Higgs boson to complete the cast of characters for the standard model of the structure of subatomic matter, seem far removed from the pressing economic concerns of most of humanity and the underlying theme of this SBE&S report — the importance of SBE&S in accelerating and sustaining U.S. economic development and technological advancement. And yet this is the same community whose previous efforts at community-scale integration and organization of scientific workflow were the origins of the World Wide Web and the Internet-based economy that we enjoy today. The current challenge faced by the particle physics community is the distillation, by seven orders of magnitude, of experimental results generated at the rate of petabytes per second, into a globally-accessible, multitier, distributed database of archival results with the associated informatics and visualization tools. Undoubtedly this unprecedented scale will break new ground for data infrastructure and knowledge management, with profound impact(s) that may perhaps rival that of the World Wide Web.

The Large Hadron Collider (LHC) (see Fig. 10.1 again) and the global experimental collaboration centered at CERN (European Center for Nuclear Research) are actually a series of interrelated experiments and detector design projects (ALICE, A Large Ion Collider Experiment; ATLAS, A Toroidal LHC Apparatus; and CMS, Compact Muon Solenoid are just three of the six detector experiments). For much of the year, the opposing beams contain protons streaming at 99.999999% of the speed of light, but about one month per year, heavy ions (e.g., lead ions) are used (see Maury). The enormity of the data challenges originates from the fact that the results of scientific interest (either proving or disproving the standard model) are extremely rare events — data architecture and the associated informatics to sift through a data torrent of ion–ion collision tracks and cross-validatation across multiple experiments are central to this grand experiment. A summary of the notable informatics tools and integration with data-intensive computing in the open science grid (OSG) is described by Livny. Finally, the enormity of the informatics challenges serves to underscore the end-to-end role of SBE&S: data intensive SBE&S is the closing book-end that pairs with the civil engineers' simulations for tunnel and instrument alignment that preceded the construction phase!

10.3 Big Data in Life Sciences Research

10.3.1 *Ecole Polytechnique Federale de Lausanne, Blue Brain Project — Switzerland*

New heights in big data are attained in this collaborative effort between EPFL and IBM to elucidate the cognitive function of the human brain with a 3D digital model ("modeling in every detail") of the cellular infrastructure and electrophysiological interactions within the cerebral neocortex. Dubbed the "blue brain project," the data challenges arise from a multiscale model featuring 10,000 neurons, 340 types of neurons, 200 types of ions and 30 million connections. The current capabilities already feature the *in silico* replication at the cellular level, the neocortical column of a rat brain. The ability to perform *in silico* replication of experimental electrical signals as a function of morphological and synaptic changes would represent a landmark advance in brain research and progress towards the cure for neurodegenerative diseases.

10.3.2 *Daresbury Laboratory/Science and Technology Facilities Council, e-HPTX — United Kingdom*

As described in our visit to Daresbury Laboratory of the STFC, the United Kingdom's fourth Collaborative Computational Project (CCP4) focused on software development for structure solution and refinement in computational and structural biology (X-ray crystallography). This e-Science and BBSRC (Biotechnology and Biological Sciences Research Council)-sponsored project recognized at a very early stage the importance of data management and workflow as foundational elements of life sciences research: the automation of structure determination, e-HPTX (management of workflow), DNA data collection, and PIMS (laboratory information management system) were all developed as part of CCP4.

10.3.3 *Systems Biology Institute — Tokyo, Japan*

The research plans at the Systems Biology Institute founded by systems biology pioneer Hiroaki Kitano include data and software infrastructure to support the SBE&S modeling efforts in systems biology. The current plans that impact on "big data" focus on the Systems Biology Markup Language (SBML, see Hucka), Systems Biology Graphical Notation (SBGNL) and Web 2.0 Biology. These elements are all crucial to an organized accumulation of systems-biological knowledge towards progress on major diseases such as cancer, diabetes, and infectious diseases. Although primarily labeled under the bigger umbrella of software infrastructure investments, the data-intensive nature of the activities in this research institute leads us to believe that this is one of the most notable and sustained investments in data cyberinfrastructure in the SBE&S landscape.

10.3.4 *Earth Simulator Center and Plans for the Life Simulator — Japan*

Jumping in scale from the molecular to the other end in the life sciences (societal scale), new big-data challenges are encountered in Japan's next-generation project, the "life simulator," featuring agent-based models with six billion agents, i.e., a one-to-one mapping to the human population on the planet (see the 2007 Earth Simulator Center Annual Report). How is the mountain of (personal!) data to be gathered and organized to simulate individual profiles and activities on such an awesome scale? What are the implications from confidentiality and privacy issues? These efforts hint at the emerging research opportunities in the social sciences, their impact on policy makers, and the inter-dependence with next generation capabilities in SBE&S as Japan transitions from the Earth Simulator to the Life Simulator. And for the world at large, this may be the first steps in applying SBE&S towards the modeling and (better!) understanding of the complex nonlinear dynamic that we are witnessing in the global financial markets, and thereby providing new tools for governmental financial authorities.

10.4 Big Data in Industry: Enterprise-Scale Knowledge Integration

As noted in the beginning of this chapter, industrial R&D laboratories with their enterprise-scale knowledge integration efforts in the cause of intellectual property (IP) generation constitute the other important group of sustained investments and activities in "big data." In light of the varying but in some cases stringent constraints on attribution of details of in-house research activities, we have aggregated without attribution the activities in big data as it pertains to workflow and knowledge management. Fortunately, such an approach has the fortuitous effect of revealing common patterns that transcend industry types, rather than constricting insights for our study.

In industrial practice, we rarely (if ever) encounter SBE&S as a "point solution" in a manufacturing process or R&D operation. Instead, SBE&S activities appear at various steps in the process or flow of work (i.e. workflow). As the work flows from step to step and functional area (department) to functional area, industrial firms face the challenge of interoperability, or how to translate the (SBE&S) output of one area to the requisite input of the next. In the chemical process industry, the CAPE-Open standard is an example of such an effort including the formation of the COLAN consortium to maintain the open standards as described by Pons (2005). The companies with leading capabilities in SBE&S have taken this problem far beyond the simplistic task of pairwise data conversion (e.g., from CAD tools to CFD simulations) and have proceeded further to implement enterprise-scale data infrastructure and workflow engines. Notable informatics capabilities include markup languages, the semantic Web (Web 2.0), and data visualization tools, e.g., in the biotechnology and pharmaceutical companies.

Some early successes of investments in data architecture, workflow, and visualization include taking weeks or months out of product development timelines and making SBE&S tools accessible to the nonexperts (companies note that no one person is an expert in all the functional areas of the workflow). We saw this trend across multiple industries including automotive, chemical, consumer products, and pharmaceutical companies. Further progress is impeded by the conflicting goals of the software vendor community and the challenges of interfacing workflow outputs with the regulatory agenices (e.g., for the biotechnology and pharmaceutical industry, see the resource constraints at the U.S. Food and Drug Administration as described in Cassell, 2007). Given the high stakes, resolution of these challenges is likely to remain a focus area for elevating the use of SBE&S to new heights in the industrial R&D setting.

10.5 Big Data — The Road Ahead for Training and Education

In this report, we have devoted significant attention to the current state of SBE&S training and education around the world (see Chapter 11). For big data, these same challenges in training and education of the next generation of SBE&S researchers are compounded by the additional complexity arising from inadequate data infrastructure. We contrast the typical state of affairs (e.g., in any major U.S. research university) and then provide a glimpse of the future by describing the best academic example (University of Stuttgart, Simulation Technology Excellence Cluster program) from our global travels.

Consider a graduate student on the university campus of a major U.S. research university today. The standard disk storage allocation for graduate students is typically 500 megabytes, or in some cases, perhaps a generous perk of 1 gigabyte (10^9 bytes) of storage. But the dissertation research may generate a terabyte (10^{12} bytes) of storage each semester! To put this in a computing perspective, this is tantamount to giving the student a portable computer instead of access to a terascale supercomputer. Fortunately, research progresses because the student runs a mini data center and infrastructure at home, using references from the Internet to build a RAID storage farm in the basement. This example underscores the gap in data infrastructure and the steps required to close that gap in terms of education and training for the typical student (i.e., someone who is not a technology enthusiast running a datacenter at home).

The Simulation Technology Excellence Cluster at the University of Stuttgart, benefits from the German Research Foundation (DFG) programs to support collaborative research centers (SBF), trans-region projects (TR), technology transfer units (TFR), research units (FOR), priority program and excellence initiatives. The result is a superbly integrated infrastructure for computation and data management as described in our site report for the university. The integrated data management and interactive visualization environment includes significant new technologies in human–computer interfaces for SBE&S-model setup and sensor networks for real-time control simulations. At Stuttgart, what we saw was not just the integration of data management and computation, but the entire spectrum of SBE&S activities — the success of this approach points to the road ahead for the global SBE&S community.

10.6 Summary of Key Findings

The role of big data and visualization in driving new capabilities in SBE&S is pronounced and critical to progress in the three thematic areas of this report. We can also interpret the "big data" challenges faced by the SBE&S community as a recursive resurfacing of the CPU-memory "data bottleneck" paradigm from the chip architecture scale to societal scale.

- The locus of large scale efforts in data management correlates by discipline and not by geographical region. The biological sciences (as described in the Chapter 2 on Life Sciences and Medicine) and the particle physics communities are pushing the envelope in large-scale data management and visualization methods. In contrast, the chemical and material science communities lag in prioritization of investments in the data infrastructure. In the biological sciences, in both academic laboratories and industrial R&D centers, there is an appreciation of the importance of integrated, community-wide infrastructure for dealing with massive amounts of data, and addressing issues of data provenance, heterogeneous data, analysis of data, and network inference from data.

- Similarly, and even within a given disciplinary area, there is a notable difference between industrial R&D centers and academic units, with the former placing significantly more attention to data management infrastructure, data supply chain and workflow, in no small part due to the role of data as the foundation for intellectual property (IP) assets. In contrast, most universities lack a campus-wide strategy for the ongoing transition to data-intensive research and there is a widening gap between the data infrastructure needs of the current generation of graduate students and the capabilities of the campus IT infrastructure.
- Industrial firms are particularly active and participate in consortia to promote open standards for data exchange — a recognition that SBE&S is not a series of point solutions but integrated set of tools that form a workflow engine. Companies in highly regulated industries, e.g., biotechnology and pharmaceutical companies, are also exploring open standards and data exchange to expedite the regulatory review processes for new products.
- Big data and visualization capabilities go hand in hand with community-wide software infrastructure; a prime example is in the particle physics community and the Large Hadron Collider infrastructure networking CERN to the entire community in a multitier fashion. Here, visualization techniques are essential given the massive amounts of data, and the rarity of events (low signal to noise ratio) in uncovering the new science.
- Big data, visualization and *dynamic* data-driven simulations are crucial technology elements in the production of transportation fuels from the last remaining giant oil fields. Global economic projections for the next two decades and recent examples of price elasticity in transportation fuels suggest that *the scale of fluctuation in reservoir valuations would be several orders of magnitude beyond the global spending in SBE&S research.*
- As was the case for the three SBE&S themes and crosscutting topics, "big data" has training and education challenges. Appropriately trained students who are adept with data infrastructure *issues will become increasingly important for research at the SBE&S frontiers.*

References

Anderson, C. (2008). The end of theory: the data deluge makes the scientific method obsolete, *Wired Magazine*, June 23.

Cassell, G. (2007). *FDA Science and Mission at Risk, Report of the Subcommittee on Science and Technology* (FDA Science Board, Washington, DC).

"Data-infrastructure-at-home" (including RAID data storage) references: http://en.wikipedia.org/wiki/RAID, www.tomshardware.com/us/, www.smallnetbuilder.com/.

Earth Simulator Center (2007). *Annual Report of the Earth Simulator Center.*

Hucka, M., *et al.* (2003). The systems biology markup language (SBML): a medium for representation and exchange of biochemical network models, *Bioinformatics*, 19, 524–531.

King, N. (2008). Saudis face hurdles in new oil drilling, *Wall Street Journal*, April 22.

Kitano, H. (2002). Systems biology: a brief overview, *Science*, 295, 1662–1664.

Livny, M. *LHC SBE&S Computational Infrastructure*, http://pages.cs.wisc.edu/~miron/.

Maury, S. *Ions for LHC Project (I-LHC)*, http://project-i-lhc.web.cern.ch/project-i-lhc/Welcome.htm.

Pons, M. (2005). *Introduction to CAPE-Open*, http://www.colan.org.

Chapter 11

EDUCATION AND TRAINING

Celeste Sagui

The country that out-computes will be the one that out-competes.
–Council on Competitiveness (http://compete.org)

11.1 Introduction

The development and improvement of computer simulations in diverse fields represent one of most important successes in science and engineering in the last 30 years. For the United States to remain competitive in the Sciences and Engineering, the proper teaching of computer modeling and simulation methods and tools in colleges and universities becomes of paramount importance. In the words of Tinsley Oden at the WTEC U.S. Baseline SBE&S Workshop (2007), "*Our educational institutions must be prepared to equip tomorrow's scientists and engineers with the foundations of modeling and simulation and the intellectual tools and background to compete in a world where simulation and the powerful predictive power and insight it can provide will be a cornerstone to many of the breakthroughs in the future.*"

The United States used to lead the way in education for simulation-based engineering and science (SBE&S). For instance, U.S. certificate programs in Computational Science and Engineering (CSE) were among the first in the world for SBE&S. However, many of these programs have not embraced new developments in the last decade. There have been many new efforts throughout the country to fill this vacuum, especially in the form of summer programs, that include schools associated with the main national codes such as AMBER, CHARMM, and NAMD. There are scattered efforts under individual PI and team NSF and DOE grants. The State of Ohio launched the Ralph Regula School of Computational Science in Ohio to focus on "blue-collar" computing, reaching out to K-12 programs and undergraduate students across the state. The Great Lakes Consortium for Petascale Computation's Virtual School of Computational Science and Engineering, launched last year to help prepare the next generation of researchers to exploit petascale computation and beyond, focuses on graduate and postgraduate education. Nanohub, which supports a virtual community primarily in nanoelectronics, offers educational components along with software access and distribution. In addition, some U.S. institutions are being restructured. For instance, the Institute for Computational Engineering and Sciences at the University of Texas, Austin, offers both education and infrastructure for interdisciplinary programs in CSE and Information Technology (IT). The faculty comes from 17 departments and four schools and colleges. The institute offers an independent PhD program, with independent faculty evaluations for promotion and tenure.

Despite these new, bold programs, the United States has seen less vigorous growth than other countries in education and training in SBE&S over the past decade. The globalization of higher education is accelerating. According to the *Science and Engineering Indicators* (NSB, 2008), the United States continues to attract the largest number and fraction of foreign students worldwide, but these numbers have generally decreased in recent years.[1] Some of the reduction in numbers of foreign students studying in the United States is believed to be the result of expanded access to higher education taking place in many countries in Europe and several countries in Asia since 1990. In addition, universities in many countries — including Australia, the United Kingdom, Canada, Japan, and Germany — have actively expanded their enrollment of foreign Science and Engineering (S&E) students. This chapter reports on the WTEC panel's findings for SBE&S education in the nation and abroad.

11.2 Where the United States Stands

Since SBE&S is a relatively new and highly interdisciplinary area of study and application, it is difficult to come up with education statistics belonging exclusively to the area. However, it is still enlightening to examine some general S&E trends for which good statistics exist.

The United States spends more on R&D than any other country in the world, approximately equal to the sum of the other G7 countries — the

[1] 2007–2008 saw an increase in international student enrollment, attributed to the faltering U.S. dollar and relaxed visa constraints for Chinese students (*World Journal*, 2008).

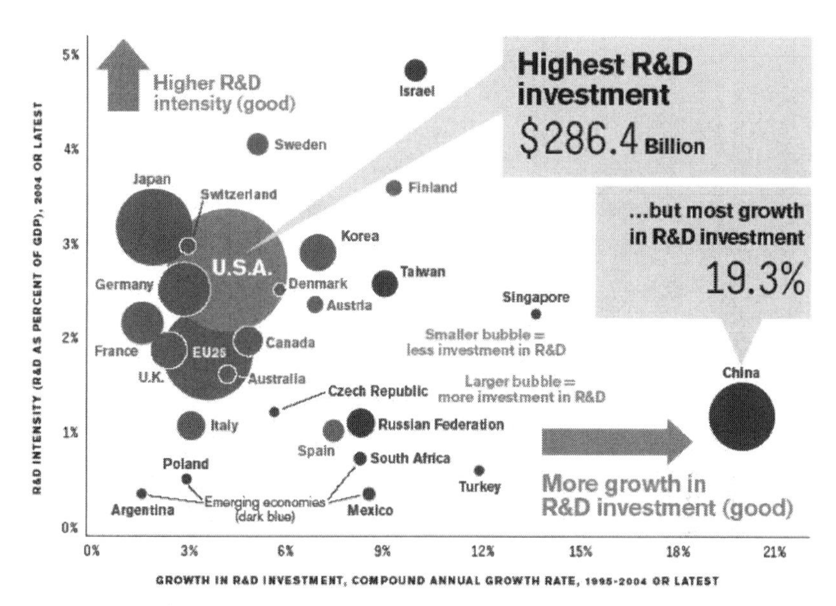

Fig. 11.1. U.S. R&D investment remains the world's largest, but others are increasing their investment faster. This figure has been taken from The Council on Competitiveness, *Competitiveness Index: Where America Stands* (2007). The figure, in turn, was created using data from OECD's *Main Science and Technology Indicators 2006* (OECD, 2006).

United Kingdom, Germany, France, Italy, Japan, and Canada — combined (NSB, 2008). Figure 11.1 gives a "map" of the countries with the highest R&D investment. The United States has more researchers (OECD, 2006) and more patents (NSB, 2006) than any other country. On the other hand, a few other countries invest more than the United States on non-defense R&D as a percentage of their GDP (e.g., Japan and Germany). At present, emerging economies (led by the People's Republic of China) have the fastest rate of growth in R&D investment.

Figure 11.2 shows that the United States has approximately 37% of global R&D spending, 52% of all new patents, 30% of all scientific publications, 29% of all scientific researchers, and 22% of all doctorates in S&E. From these figures, it is clear that the United States continues to be an absolute leader in S&E. However, Fig. 11.2 also shows that its global share has fallen as other countries have increased their science and technology initiatives and efforts.

Although the United States still has the most citations and the top-cited publications, for the first time the European Union has surpassed the

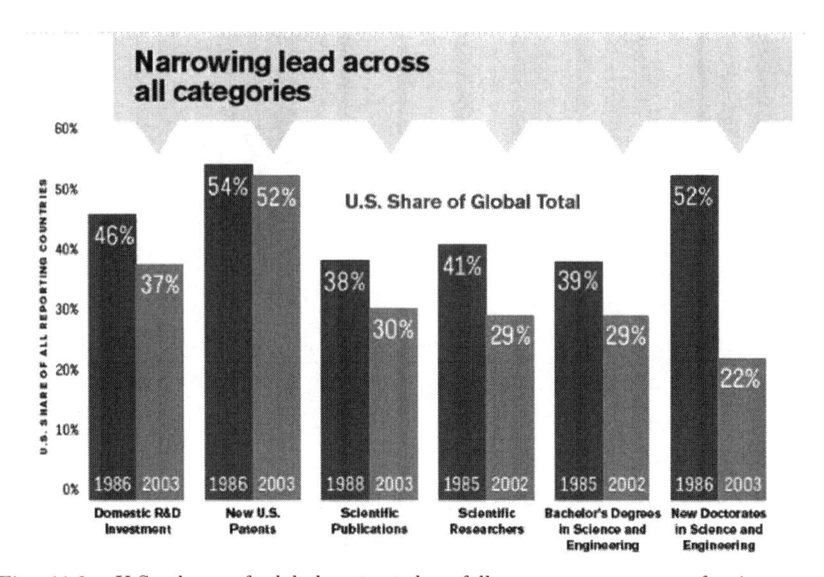

Fig. 11.2. U.S. share of global output has fallen across a range of science and technology metrics. This figure has been taken from The Council on Competitiveness, *Competitiveness Index: Where America Stands* (2007). The figure was created using data from OECD's *Main Science and Technology Indicators* (2006), NSF's *Science and Engineering Indicators* (NSB, 2006), and the U.S. Patent and Trademark Office.

United States in total number of scientific publications and the production of PhDs in S&E, according to the data presented in Figs. 11.3 and 11.4. In particular, the WTEC bibliometrics study (see Appendix C) shows that the number of SBE&S publications worldwide is double the number of all S&E publications (5% versus 2.5%), and while in 2007 the United States dominated the world SBE&S output at 27%, China moved up to second place, at 13% (although with low citation indexes). In addition, the U.S. output in SBE&S is less than that of EUR-12 (12 countries of the European Union), with that difference increasing over time.

Problems that especially affect the SBE&S fields include the following:

(1) Loss of international students: this is due to both visa restrictions and the diminution of the country's "attractiveness" to foreigners.
(2) Loss of PhD students and faculty: the scientific field is no longer attractive to U.S. students; faculty and principal investigators and staff researchers in national labs are being lost to industry and other markets.

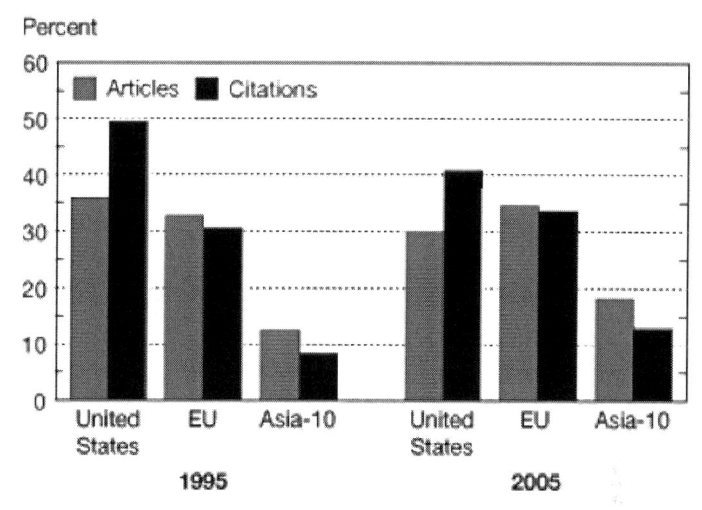

Fig. 11.3. S&E articles and citations in all fields, by selected region/country: 1995 and 2005. The figure is extracted from *Science and Engineering Indicators* 2008 (NSB, 2008). The share of all articles is based on a 3-year period. Article counts are from a set of journals covered by the Science Citation Index (SCI) and Social Sciences Citation Index (SSCI). Articles are classified by the year they entered the database and are assigned to region/country/economy on basis of institutional addresses listed in the article. In the case of articles and citations on a fractional-count basis, i.e., for articles with collaborating institutions from multiple countries/economies, each country/economy receives fractional credit on the basis of the proportion of its participating institution. Citation data are based on the year the article entered the database. Citation counts are based on a 3-year period with 2-year lag, e.g., citations for 1995 are references made in articles in the 1995 data tape to articles in the 1991–1993 data tapes.

(3) Lack of adequate training of students in SBE&S: students are knowledgeable in running existing codes (with visualization), but they are unable to actually write code or formulate a mathematical framework.
(4) Rigid "silo" academic structure: the entire academic structure (budgets, courses, curricula, promotion, tenure, etc.) is aimed at maintaining a vertically structured, relatively insulated disciplinary environment. The system promotes departmental loyalty and highly discourages interdisciplinary work.
(5) Serious danger of compromising creativity and innovation: in the present funding environment, many alliances are made for the purpose of seeking funding, independent of the scientific value of the research.

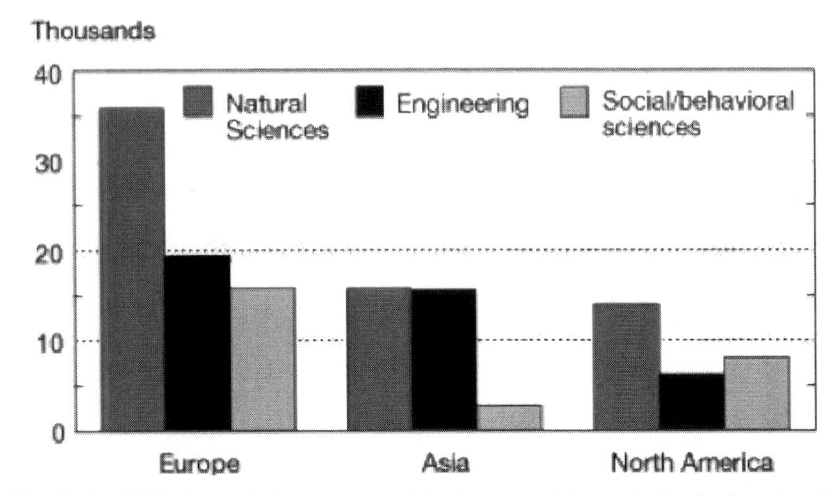

Fig. 11.4. S&E doctoral degrees earned in Europe, Asia, and North America, by field: 2004 or most recent year (before, 2004). The figure is extracted from *Science and Engineering Indicators 2008* (NSB 2008). Natural Sciences include physical, biological, earth, atmospheric, ocean, agricultural, computer sciences, and mathematics. Asia includes China, India, Japan, South Korea, and Taiwan. Europe includes Western, Central, and Eastern Europe. North America includes the United States and Canada. This figure was compiled based on the following sources: Organization for Economic Co-operation and Development (2008), Education Online Database; United Nations Educational, Scientific, and Cultural Organization (UNESCO) (2007), Institute for Statistics database, http://www.unesco.org/statistics (April, 2007); and national sources.

The U.S. loss of international students and postdocs due to restrictions on visas is far from breaking news. As pointed out by an article in Lexington, 2008, Help Not Wanted, April 10, 2008:

> "This [the restriction on visas] is a policy of national self-sabotage. America has always thrived by attracting talent from the world. Some 70 or so of the 300 Americans who have won Nobel prizes since 1901 were immigrants (and many others were direct descendants from immigrants). Great American companies such as Sun Microsystems, Intel and Google had immigrants among their founders. Immigrants continue to make an outsized contribution to the American economy. About a quarter of information technology (IT) firms in Silicon Valley were founded by Chinese and Indians. Some 40% of American PhDs in science and engineering go to immigrants. A similar proportion of all the patents filed in America are filed by foreigners."
>
> "These bright foreigners bring benefits to the whole of society. The foreigner-friendly IT sector has accounted for more than half of

America's overall productivity growth since 1995. Foreigner-friendly universities and hospitals have been responsible for saving countless American cities from collapse. Bill Gates calculates, and respectable economists agree, that every foreigner who is given an H1B visa creates jobs for five regular Americans."

Other countries are benefiting enormously from the "leak" of talent that is affecting the United States. For instance, instead of the lottery system employed by the United States for assigning H-1B (temporary resident/foreign guest worker) visas, countries like Canada and Australia have a merit system that rewards educational accomplishments. Some companies in New Zealand hand out work visas with their job offers. The European Union is also considering a merit system to allow talented people to become European Union citizens.

Figure 11.5 shows first-time graduate enrollment in Science and Engineering in the United States by citizenship and field from 2002 to 2006. Visa holders constitute approximately 50% of the total S&E enrollment, surpassing in 2004–2006 the number of U.S. citizens and permanent residents enrolled in Computer Sciences and Engineering. Although other countries (such as the United Kingdom) also have a very large proportion of international students, the absolute number of international students in the United States exceeds by far those from other countries.

Figure 11.6 shows the absolute number of international students per country (left) and the percentage of doctorate degrees earned by foreigners in the United States, the United Kingdom, Germany, and Japan.

According to the most recent *Science and Engineering Indicators* (NSB, 2008), students on temporary visas earned more than a third (36%) of all S&E doctorates awarded in the United States in 2005. Temporary residents earned half or more of all U.S. doctorates in Engineering, Mathematics, Computer Sciences, Physics, and Economics in 2005.

The number of temporary resident postdoctoral students in the United States has definitely surpassed the number of U.S. citizen/permanent resident postdoctoral students in S&E (Fig. 11.7). The temporary resident numbers among postdoctoral students went from 8859 in 1985 to 26,975 in 2005, while the number of U.S. citizens/permanent residents among postdoctoral students went from 13,528 in 1985 to 21,678 in 2005, so that temporary visa holders accounted for 55% of all S&E postdocs in academic institutions in the fall of 2005. In particular, the percentage of temporary visa holders in 2005 was 59% in the Biological Sciences, 60%

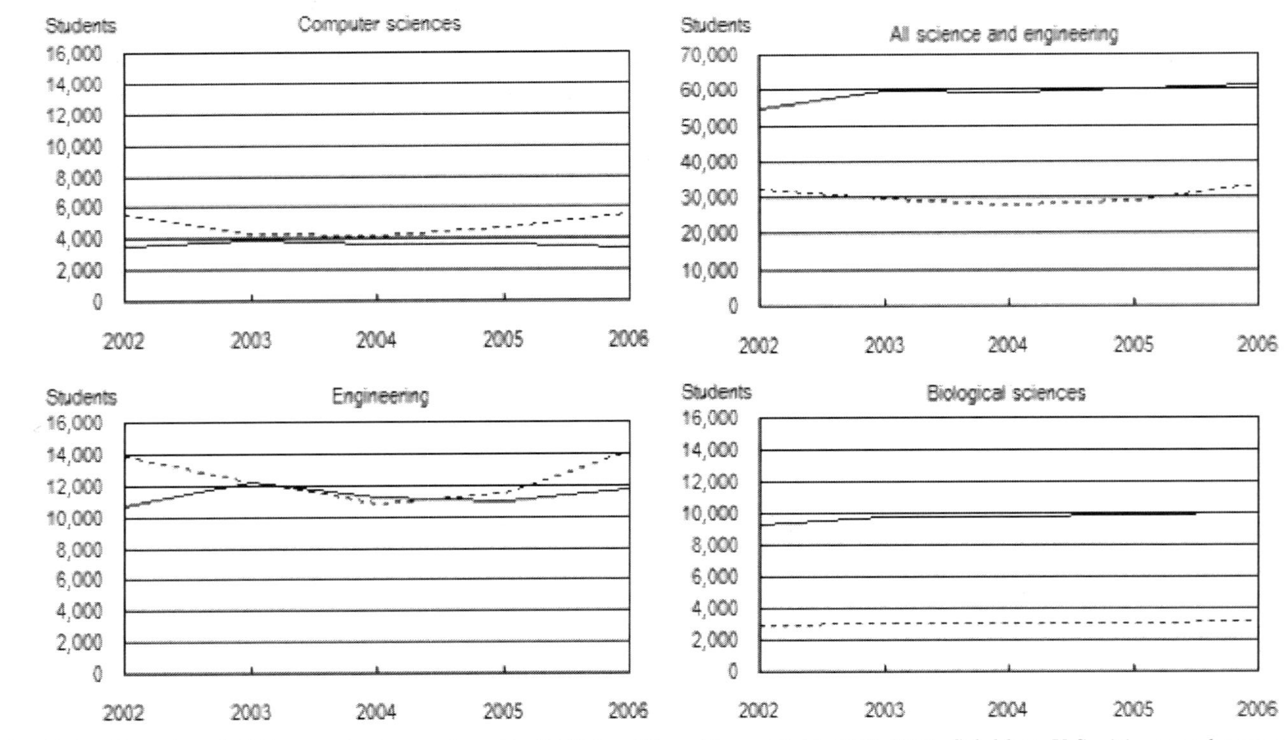

Fig. 11.5. First-time, full-time graduate enrollment in S&E, by citizenship and field, 2002–2006. *Solid line*, U.S. citizens and permanent residents; *dotted line*, visa holders (NSF, 2007).

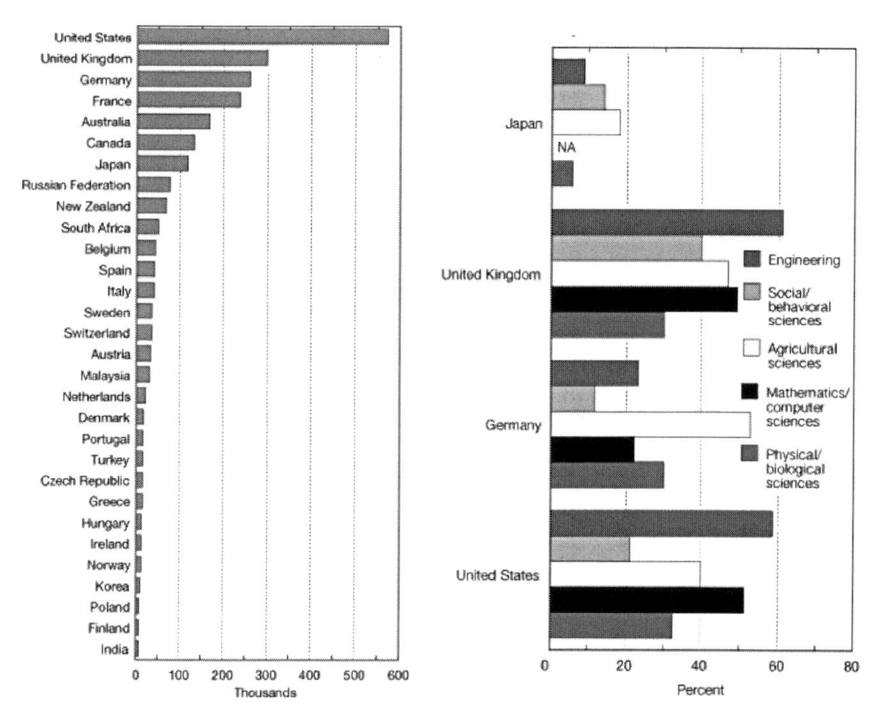

Fig. 11.6. *Left*: foreign students enrolled in tertiary education, by country: 2004. *Right*: S&E doctoral degrees earned by foreign students, by selected industrialized country and field: 2005 or most recent year (before, 2005). Both figures are obtained from *Science and Engineering Indicators 2008* (NSB, 2008).

in Computer Sciences, 66% in Engineering, and 64% in the Physical Sciences (NSB, 2008).

In the United States, the *total* number of S&E doctoral degrees in Engineering, Mathematics, Computer Sciences, and Physical Sciences started falling around 1995–1996 and only began to regain the 1995–1996 values a decade later (NSB, 2008). When it comes to *relative* (percentage) performance, the United States performs worse than other countries in some areas of S&E. Figure 11.8 shows the academic R&D share of all R&D for selected countries and economies. The United States lags behind several emerging and developed economies. The right panel shows the percentage of 24-year-olds holding Natural Sciences or Engineering degree by country or economy.

Other alarming statistics obtained from *Rising above the Gathering Storm: Energizing and Employing America for a Brighter Economic Future*

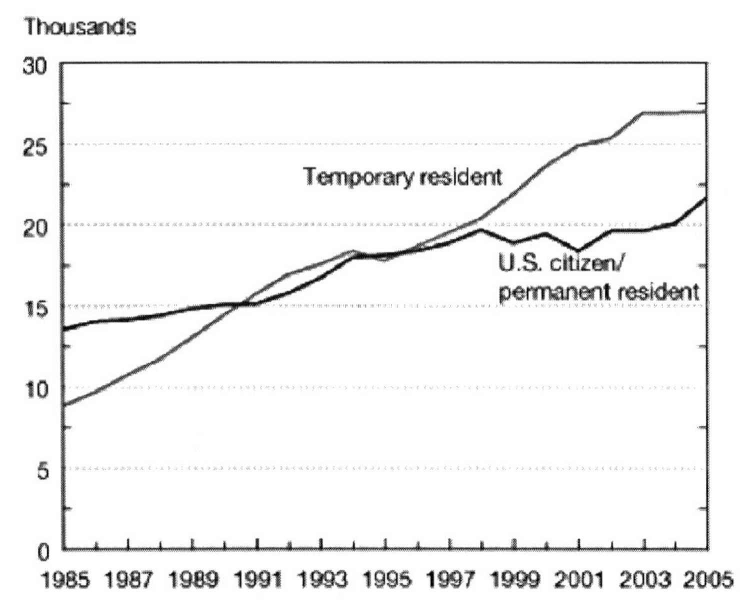

Fig. 11.7. S&E postdoctoral students at U.S. universities, by citizenship status: 1985–2005. Figure is taken from *Science and Engineering Indicators 2008* (NSB, 2008).

(COSEPUP, 2007 and references therein) are the following. The percentage of U.S. undergraduate students with an S&E degree is relatively low: South Korea, 38%; France, 47%; China, 50%; Singapore, 67%; United States, 15%. In the year 2000, an estimated 38% of the total U.S. science and technology workforce was foreign-born. About one-third of U.S. students intending to major in Engineering switch majors before graduation. More chief executive officers of Standard & Poor's 500 companies obtained their undergraduate degree in Engineering than in any other field. There were almost twice as many U.S. Physics BSc degrees in 1956 than in 2004. Federal funding of research in the Physical Sciences, as a percentage of GDP, was 45% less in the fiscal year (FY) 2004 than in FY1976.

11.3 How Other Countries Compare

Faculty and researchers everywhere are more interested in computing and open to the myriad of computing possibilities than ever before. This is due to the increased awareness of simulation successes, important changes in industrial attitude, and the demand for qualified computational scientists

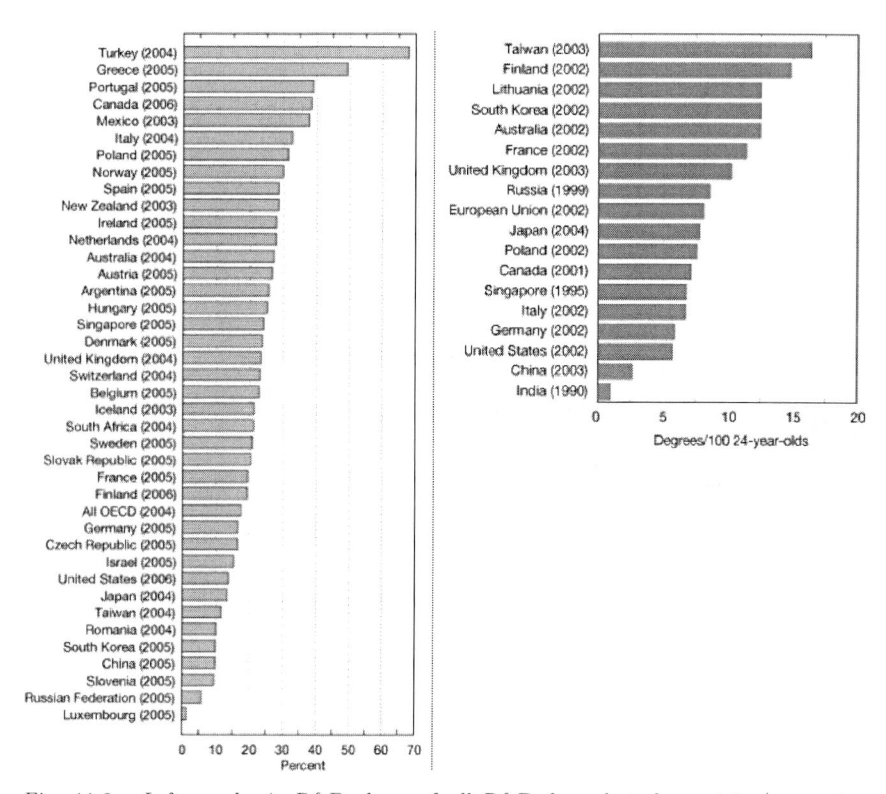

Fig. 11.8. *Left*: academic R&D share of all R&D for selected countries/economies and all OECD. Figure is taken from *Science and Engineering Indicators 2008* (NSB, 2008). *Right*: Natural Sciences and Engineering degrees per 100 24-year-olds, by country/economy. Figure is taken from *Science and Engineering Indicators 2006* (NSB, 2006).

in all areas of R&D. Virtual experimentation, with its increased capabilities for explanation and prediction, is rapidly gaining ground across a wide spectrum of industrial and academic activities. Students are attracted to SBE&S because they can easily get good jobs afterwards. Here, we report on some general trends that the WTEC panelists witnessed in our site visits. For further details, references, etc., the reader is referred to the site reports on the International Assessment of Research and Development in Simulation-Based Engineering and Science website (http://www.wtec.org/private/sbes/. The required password is available from WTEC upon request).

11.3.1 *Finding 1: There is Increasing Asian and European Leadership in SBE&S Education due to Dedicated Funding Allocation and Industrial Participation*

The United States still provides strong public investment in R&D, but other countries are also taking important initiatives to provide funding for education and to foster industrial collaboration in order to earn an edge in the global SBE&S innovation environment. This trend is clearly seen in Fig. 11.2 and also in the somewhat alarming results presented in Fig. 11.8 (left) that show the academic share of all R&D for the United States lagging behind that of several important competitors. Although specific statistics for SBE&S do not exist, it is the impression of this panel, as analyzed in various chapters, that the U.S. leadership in application software development in SBE&S is decreasing, due in part to relatively less funding.

Most software development these days takes place at universities and national laboratories. This is the case both in the United States and abroad. Worldwide, industry and private companies find it too onerous to invest in software development within their own working habitat and tend to base their simulation activities on software developed somewhere else. Of course, companies do invest in software when their own specific needs are not met by existing codes, and Chapter 10 provides several examples of this situation, along with a more detailed discussion. The WTEC panel found that the way companies are supporting SBE&S research is by funding intra- and extramural research at university groups. Examples of this are found in the Institute of Process Engineering in the Chinese Academy of Science, the Department of Engineering Mechanics at Tsinghua University in China, the Center of Biological Sequence in Denmark, CIMNE in Barcelona, and many other institutions. Germany is a particularly striking example of corporate investment, where the share of academic R&D financed by industry considerably exceeds that of any other country (NSB, 2008).

Interesting examples of Asian institutions that benefit from healthy government/industry funding in SBE&S are the following:

- *Japan Earth Simulation Center (ESC).* The primary objective of the ESC is to develop new algorithms, with a focus on multiscale and multiphysics algorithms. It offers industrial participation with a fee structure controlled by the Japanese government. The ultimate goal of the ESC is to simulate physical systems as realistically as possible. One major ESC success is global climate simulation with resolutions of 10 km, and resulting improvement in weather prediction capabilities. The ESC is scheduled to be closed, and the Japanese government is investing heavily in software development to be run on the next supercomputer that will replace the ESC. This is being done in spite of the fact that it not known what form the new supercomputer will take. The goal is that innovation in algorithms will drive at least part of the hardware.

- *The Systems Biology Institute (SBI, Japan).* SBI was founded by Dr. Hiroaki Kitano, one of the early pioneers of Systems Biology. The institute, which also has experimental labs in the Cancer Institute of the Japan Foundation for Cancer Research, has been funded by the Japanese government for 10 years, including through a grant for international software standards formation. The current research plan focuses on the development of experimental data and software infrastructure. The latter includes Systems Biology Markup Language (SBML), Systems Biology Graphical Notation (SBGN), CellDesigner, and Web 2.0 Biology, designed for the systematic accumulation of biological knowledge. Dr. Kitano believes that it is crucial to invest in software infrastructure because software is critical to the development of systems biology as a field. In recognition that it is difficult to publish software, the merit system in this lab values software contributions as well as publications.
- *University of Tokyo.* Ranked number one in Japan, the University of Tokyo is one of the recipients of the "21st Century Center of Excellence (COE) Program" awards. This program, created by the Ministry of Education, Culture, Sports, Science and Technology, aims to form world-class research and education centers at the universities in Japan; 28 of these centers were established at the University of Tokyo. Many of the original five-year COE projects have ended, but some of these have become part of the Global COE program, which aims to form international research and education excellence programs. The program in Mechanical Systems Innovation directed by Professor Kasagi is highly interdisciplinary, has three main focus areas — energy, biomedicine, and hypermodeling/simulation — and includes a big education component.
- *Institute of Process Engineering (IPE), Chinese Academy of Sciences.* About 50% of the research funding for IPE comes from industrial sponsors, both domestic and international, with significant funding from the petrochemical industry. IPE also receives significant government funding through the National Natural Science Foundation of China

and the Ministry of Science and Technology. Its main SBE&S focus is multiscale simulations for multiphase reactors. In addition to commercial packages, IPE researchers have developed their own codes for particle fluidization and reaction kinetics. Students are especially trained on the use of these codes and packages.

- *Tsinghua University Department of Engineering Mechanics (China).* Tsinghua's Engineering Mechanics department is believed to be the largest in the world, and it is ranked number one in China. The department has several high-profile SBE&S collaborations with multinational companies. Of the total annual research budget, 50% comes from the National Natural Science Foundation of China (NSFC), and 30% is from industry-sponsored projects. Even though training in SBE&S-related activities is highly valued by the students because of resulting career opportunities, the researchers regret that the NSFC does not fund code development. However industry often steps into this gap (for instance, simulations of cardiac muscles were sponsored by Medtronic, Inc.).
- *Fudan University, Shanghai (China).* The work of the Department of Macromolecular Science and the Institute of Macromolecular Science at Fudan University on computational polymer physics is on par with that of leading groups in the United States, UK, Germany, etc. There is strong emphasis on education. Training on analytical work is required prior to training in computational work, to avoid producing students that view simulation software packages merely as "black boxes." Prof. Yang is both the Director of the Computational Polymer Physics Group and the Vice Minister of Education. In this position, he oversees the "211" program in Beijing, whose aim is to solve problems important for China, to increase the number and impact factors of China's publications, and to elevate the quality of PhD theses. Strong funding has been allocated for SBE&S and to support 2000 graduate students per year to study abroad.
- *King Abdullah University of Science and Technology (KAUST).* This university is being built in Saudi Arabia as an international, graduate-level research university with the highest endowment-per-student worldwide. KAUST, scheduled to open in September 2009, is supported by a multibillion dollar endowment, and governed by an independent Board of Trustees. It intends to attract students from around the world, with main research thrusts in (i) resources, energy and environment; (ii) biosciences and bioengineering; (iii) materials science and engineering; and (iv) applied mathematics and computational sciences.

These thrusts will be pursued in multidisciplinary Research Centers where SBE&S plays a major role. Recently, KAUST and IBM announced the creation of a research partnership to build and conduct research on the most complex, high-performance computing system in the region and among academic institutions in the world. In a joint statement, they said the new system, named Shaheen, will serve the university's scientific researchers and put the university on a path to exascale computing in the near future (KAUST, 2008). KAUST has also announced a partnership with UC San Diego to build the world's most advanced visualization center, named Cornea. The Geometric Modeling and Scientific Visualization Research Center (GM&SVRC) will allow researchers to transform raw data into a fully three-dimensional visual experience and will support KAUST's ambitious plan to deploy state-of-the-art technologies for elite scientific research (*Primeur Monthly*, 2008).

Signs of increasing European leadership due to increased funding in SBE&S were observed across a large number of institutions. Some examples follow:

- *Center for Biological Sequence Analysis (CBS, Denmark).* CBS, part of the Bio-Centrum at the Technical University of Denmark (DTU), is one of the largest bioinformatics centers in the European Union. Its research is focused on the functional aspects of complex biological mechanisms. CBS is funded — in addition to a contribution by DTU — by the Danish Research Foundation, the Danish Center for Scientific Computing, the Villum Kann Rasmussen Foundation and the Novo Nordisk Foundation (U.S.$100 million), as well as from other institutions in the European Union, industry, and the U.S. National Institutes of Health (bioinformatics, systems biology). CBS success is measured by its large number of publications in high-profile journals (*Science*, *Nature*, *Molecular Cell*, etc.) and its very high citation records.
- *CIMNE, the International Center for Numerical Methods in Engineering (Barcelona, Spain).* CIMNE is an autonomous research center created in 1987 by the Polytechnic University of Catalonia, the government of Catalonia, and the government of Spain, under the auspices of the United Nations Education Science and Cultural Organization (UNESCO). CIMNE's mission is three-fold: research, training, and technology transfer, all in the SBE&S area. More than 90% of its funding

comes from external sources, with an annual funding of €10 million. CIMNE has strong links to industry, with about 50% of its research funded by industry: aerospace, marine, civil, mechanical, biomedical, and metal companies. CIMNE strongly competes for European Commission projects, with a high success rate, and is a shareholder in three companies specializing in marine engineering, civil and structural engineering, and aeronautics and space applications.

- *Germany.* German SBE&S has undergone spectacular growth due to both government and industry funding. The German Research Foundation (DFG) has provided support for collaborative research centers (SFB), transregional collaborative research centers (TRR), transfer units (TBF), research units (FOR), priority programs (SPP), and "Excellence Initiatives." Many of these are based on or have major components in SBE&S. The reader is referred to the site reports in the Europe Site Reports section of the International Assessment of Research and Development in Simulation-Based Engineering and Science website, http://www.wtec.org/private/sbes/ (required password available from WTEC upon request) on the Technical University of Munich, the University of Karlsruhe, and the University of Stuttgart (which will be further discussed below), for a better picture of Germany's scientific "renaissance."
- *Fraunhofer Institute for the Mechanics of Materials (IWM) (Germany).* The Fraunhofer Institutes comprise the largest applied research organization in Europe, with an annual research budget of €1.3 billion and 12,000 employees in 56 institutes. The IWM has a €15.5 million annual budget, with 44% of the budget coming from industry and another 25–30% from the government. It is crucial for the funding model of the IWM that its base funding is a fixed percentage of the industrial funding. The IWM has seen significant growth in recent years (10% per year). At the IWM, fully 50% of the funding supports modeling and simulation (a number that has grown from 30% five years ago). The IWM has a world-class effort in modeling of materials (micromechanical models for deformation and failure/fracture, damage analysis, etc.); simulation of manufacturing processes (pressing, sintering, forging, rolling, reshaping, welding, cutting, etc.); and simulation of components (behavior, upper limits, lifetime, visual testing, etc.). It has a novel method for developing students into potential hires based on awarding competitive €50,000 projects to fresh PhDs to work at IWM in the topics of their choice.

11.3.2 *Finding 2: There are a Number of New EU Centers and Programs for Education and Training in SBE&S — All of them of an Interdisciplinary Nature*

Here are just a few examples of the many new centers and programs that have been created in Europe:

- *CBS (Denmark).* The CBS offers an MSc in Systems Biology and another in Bioinformatics that are loosely structured, not linked to any particular department or school. CBS offers realtime Internet training (all lectures, exercises, and exams in cyberspace). Typically, half the students participate onsite and half participate over the Internet. International exchange is highly encouraged; students can take their salary and move anywhere on the globe for half a year.
- *CIMNE (Spain).* CIMNE organizes courses and seminars on the theory and application of numerical methods in engineering. The attendees are recent university graduates and also professionals. In the last 20 years, CIMNE has organized 100 courses, 300 seminars, and 80 national and international conferences, and has published 101 books, 15 educational software packages, and hundreds of research and technical reports and journal papers.
- *ETH Zürich (Switzerland).* ETHZ offers campus-wide programs in Computational Science and Engineering, which include the Computational Collaborative, the Institute of Computational Science, and the CSE degree program. The Computational Collaborative provides an umbrella for research across different SBE&S areas and has been very popular with postdocs. The Institute of Computational Science entails interdisciplinary research: Visual Computing, Bioinformatics, Computational Biology, Marine Learning and Bio-inspired Computation, Parallel Computing, Large Scale Linear Algebra, Agent Based Modeling and Simulation, Advanced Symbolic Computation, Multiscale Modeling and Simulation, Systems Biology, Virtual and Augmented Reality, Human-Computer Interface, Pattern Recognition, Nanotechnology, Art, Entertainment, and Finance. The CSE degree program is a pioneering program in Europe offering both BSc and MSc degrees in interdisciplinary areas and combining several departments. The program is also popular with graduate and postdoctoral students who take the senior-level course.
- *Technische Universität München (TUM) and Leibniz Supercomputing Center (LRZ) (Germany).* TUM offers a wealth of computational

programs: (1) the Bavarian Graduate School of Computational Engineering (BGCE) offers a Bavaria-wide MSc honors program (an "elite" program); (2) the International Graduate School of Science and Engineering (IGSSE) offers a PhD-level program that is part of the German Excellence Initiative; (3) the Center for Simulation Technology in Engineering provides research training for PhD students; (4) the Centre for Computational and Visual Data Exploration provides research training for PhD students; (5) there are regular MSc programs in traditional areas of Informatics and also in Mathematics for Engineering, Computational Physics, Computational Mechanics, and CSE; (6) there are two international (in English) Master's programs in Computational Mechanics and Computational Science and Engineering, with multidisciplinary cooperation involving 7 departments (they also allow for industrial internships); (7) a working group of CSE programs in Germany/Austria/Switzerland (München, Stuttgart, Darmstadt, Frankfurt, Aachen, Braunschweig, Rostock, Erlangen, Bochum, Bremen, Hannover, Dresden, Zürich, Basel, and Graz) that was established in 2005 obtained the legal status of an "eingetragener Verein" (registered voluntary association) in 2008; in addition, TUM has many other programs with other universities and industry. An innovative effort within the BCGE is the Software Project, which promotes the development of software for high-performance computing and computational science and engineering as an educational goal.

11.3.3 *Finding 3: EU and Asian Education/Research Centers are Attracting an Increasing Number of International Students from All Over the World, Including the United States*

Examples of this trend in Asia are as follows:

- *Japan.* The main target of the University of Tokyo's Global COE Program is support for international education and research. The National Institute for Materials Science (NIMS) recently created the International Center for Young Scientists, based on four premises: (1) completely international, with English as the language of instruction; (2) independent and autonomous research; and work that is (3) interdisciplinary and (4) innovative. Benefits to participants include a high salary, research grant support, and an ideal research environment.

Japan is investing heavily in recruiting foreign students. Slightly below 120,000 overseas students were enrolled in Japanese universities in 2007 and the Prime Minister wants to raise that number to 300,000. The situation in Japan is further described in Finding 7 below.

- *China.* An important aim of China's "211" and "985" programs is to build world-class universities that will also attract foreign researchers. The number of foreign students has increased quite dramatically in China. In 1950, the People's Republic of China received its first group of 33 students from East European countries. By the end of 2000, the total number of international students (1950–2000) in China was 407,000 (Ministry of Education of the People's Republic of China, 2000). A total of 195,503 international students from 188 countries and regions came to study in China in 2007. They were distributed over China's 31 provinces, autonomous regions, and municipalities (excluding Taiwan Province, Hong Kong and Macao Special Administrative Region), and enrolled in 544 universities and colleges, scientific research institutes, and other teaching institutions. China's top five student source countries were the Republic of Korea (64,481 students), Japan (18,640), the United States (14,758, mainly Chinese-American), Vietnam (9702), and Thailand (7306) (*People's Daily Online*, 2008). According to the Observatory on Borderless Higher Education, Malaysia and Singapore are also seeking to attract more foreign students by building up their university systems and offering more programs in English (Jaschik, 2007).
- *King Abdullah University of Science and Technology (KAUST).* Saudi Arabia is aflush with money and this is reflected in the creation of KAUST, as mentioned in Finding 1. KAUST, the KAUST-IBM Center for Deep Computing, and the KAUST-UCSD Center for Geometric Modeling and Scientific Visualization are recruiting computational scientists and engineers at all levels, including, of course full-fellowship Master's and Doctoral students. KAUST is perfectly positioned to attract the best and brightest students and researchers from the Middle East, China and India.
- *Australia.* Australia has heavily targeted students in Malaysia and Taiwan for recruitment to its universities, leading to increases in enrollments in recent years. At the same time, the United States and the UK have seen their number of Taiwanese (U.S.) and Malaysian (UK) enrollments decrease (Jaschik, 2007).

Examples of this trend in Europe are as follows:

- *United Kingdom.* As shown in Fig. 11.6 (left), the United Kingdom ranks second in the world in attracting international students. In all the different university departments that the WTEC panel visited in England (University College London, University of Cambridge, University of Oxford), hosts commented on the number and excellent quality of international students. The reputation of these universities gives them their pick of the best and brightest students and postdocs.
- *CBS (Denmark).* The CBS Internet courses are used to attract international students. They require 20% more effort than traditional courses but bring in lots of revenue and are always oversubscribed.
- *CIMNE (Spain).* (1) CIMNE introduced an international Master's course (Erasmus Mundus Master Course) in computational mechanics for non-European students; it started in 2008 with 30 students. There are four universities involved in this course (Barcelona, Stuttgart, Swansea, and Nantes). (2) CIMNE also has a Web environment for distance learning, where it hosts a Master's course in Numerical Methods in Engineering and other postgraduate courses. (3) It manages a collaborative network of "CIMNE Classrooms," physical spaces for cooperation in education, research, and technology located in Barcelona, Spain, Mexico, Argentina, Colombia, Cuba, Chile, Brazil, Venezuela, and Iran.
- *ETH Zürich (Switzerland).* The number of international students at ETHZ has increased dramatically. Now there are many excellent Asian students and Russian students who are first-rate in mathematical skills.
- *Vrije University Amsterdam (The Netherlands).* Half of the Vrije graduate students come from outside the Netherlands, mainly from Eastern Europe. In Theoretical Chemistry, the strongest and most disciplined students come from Germany.
- *TUM (Germany).* In addition to TUM's international programs mentioned in Finding 2 (international MSc, IGSSE, regional collaborations), it has many other activities on an international scale. It established Computational Engineering in Belgrade, Computational Science in Tashkent/Uzbekistan, Applied and Computational Physics in St. Petersburg/Russia, and it has planned a joint CSE program with the National University of Singapore. Simulation is a very internationally oriented field, and many of TUM's partnerships have their origin in simulation collaborations. About 80% of the SBE&S students in MSc

programs come from abroad: the Middle East, Asia, Eastern Europe, and Central and South America.

- *DTU Wind Engineering (Denmark).* DTU is one of the very few places in the world to offer both MSc and PhD degrees in Wind Engineering. The DTU MSc international program attracts students from around the world.

11.3.4 *Finding 4: There are Pitfalls Associated with Interdisciplinary Education: Breadth Versus Depth*

An issue that educators and researchers seem to be grappling with is the inescapable fact that educational breadth comes at the expense of educational depth. There is a general feeling that Computer Science students can spend too much time on the "format" of a computer program without really understanding the underlying science. Thus, many faculty and researchers prefer to hire students with a relevant science background (generally Physics or Chemistry) when dealing with research issues that need to find expression in a software program. This view was expressed in many different places: CRIEPI in Japan, ETHZ in Switzerland, and others.

Some groups have come up with an "ideal background" for a student. For instance, Dr. Kitano in the Systems Biology Institute in Japan believes that the ideal educational background for students in Systems Biology may be an undergraduate degree in Physics, Master's degree in Computer Science, and PhD degree in Molecular Biology. ETH Zürich has had success with postdocs having backgrounds in both Science/Engineering and Computing Sciences. Many researchers have noted that in order to solve "grand challenge" problems in a given field, solid knowledge of a core discipline, in addition to computational skills, is absolutely crucial.

Of course, another ubiquitous problem of interdisciplinary research is appropriate evaluation of scientific performance. At present, this is generally carried out within the confines of disciplinary indicators, and it is challenging to come up with a system of credit attribution in interdisciplinary endeavors. Moreover, as pointed out by a participant in the panel of TUM/LRZ in Munich, there is the question of the "hidden innovation phenomenon:" *the prize/profit/recognition goes to the scientist or engineer who runs a given code and not to the person(s) who write(s) the code that enables the scientific breakthrough.*

11.3.5 *Finding 5: Demand Exceeds Supply: Academia Versus Industry*

There is huge demand for qualified SBE&S students, who get hired immediately after earning their MSc degrees and often do not even go into PhD programs. Students go to different industrial settings: to pharmaceutical, chemical, oil, (micro)electronics, IT, communications, and software companies; to automotive and aerospace engineering; or to finance, insurance, environmental institutions, and so forth. This is good insofar as it maintains a dynamic market workforce, but academia would like to see more students continue a tradition of unattached, basic research.

11.3.6 *Finding 6: There is Widespread Difficulty Finding Students and Postdocs Qualified in Algorithmic and Software Development*

An important distinction between software and middleware was emphasized in Chapter 10. Because of the U.S. strength in hardware development, the United States leads the world in system-level tools and middleware. However, the development of open source systems and tools has made system-level software and middleware universally available. Therefore, most countries are not competing for leadership in these areas but for leadership in applications software, where the highest impact on science and engineering is expected to take place.

Unfortunately, postdocs and research staff with appropriate training in the development of algorithms and programs are hard to recruit. Most students in United States and abroad are trained primarily to run existing (generally commercial) codes to solve applied problems rather than learning the skills necessary to develop new codes. Many of the WTEC panel's hosts felt that a big hurdle in programming is insufficient math and physics preparation; another common complaint is that the students are not critical about the results they obtain from a simulation and often do not even know the physical approximations inherent to the codes or underlying algorithms, or the range of validity of these approximations. These issues, in some form or other, were almost ubiquitously expressed during the WTEC panel's site visits: at RICS in Japan, Vrije University in Amsterdam, the Theory of Condensed Matter Group at Cambridge University, IBM Zürich, ETH Zürich, Imperial College London, University College London, the NIMS Computational Material Science Center (CMSC) in Japan, the Institute of

Chemistry of the Chinese Academy of Sciences, the Fraunhofer Institute for the Mechanics of Materials in Germany, and others.

Some institutions or groups (for instance, Dalian University of Technology and the Peking University Center for CSE in China) insist that their students develop their own software, for educational purposes, even if it means reinventing the wheel. German institutions seem to have taken a more aggressive approach. For instance, the Technical University of Munich, University of Karlsruhe, and University of Stuttgart all have emphasized software development thorough such means as their elite programs, degrees in SBE&S, and productive participation in the supercomputer centers.

Indeed, the different SBE&S curricula should always put emphasis on proper software engineering, and — as suggested by the panel's hosts at ETH Zürich and other institutions — development of open source code should be encouraged, as it generally leads to more carefully designed software to which other researchers can contribute. Unfortunately, maintenance of big codes is always a problem in academia, especially when it comes to obtaining the funding for such maintenance.

Although many of the problems described above affect almost every country, it is still eye-opening to the panel to see that U.S. leadership in applications software development for SBE&S has decreased considerably in the last decade, during which time the centers of algorithm development shifted to Europe. This situation, described more quantitatively in Chapters 9 and 10, is partly attributed to the expectation of U.S. funding agencies and university administrators for high-visibility science results that are published in high-profile journals. Software engineering is a long-term endeavor, and such expectations undermine support for long-term progress. Moreover, the pressure of tenure and promotion in the United States actively discourages research that takes a long time to see through to publication. When a faculty member finally reaches a senior level, and in principle is free from these pressures, he or she has to consider the future careers of students/postdocs, again undermining software development in favor of faster accomplishments.

11.3.7 *Finding 7: Population Matters*

One other factor that affects not only SBE&S but also the entire spectrum of Sciences and Engineering R&D (and indeed, the entire economy of a country) is population. Often disregarded in science (because many of the most populated countries tend to be the least developed scientifically), the reality brought about by dwindling populations in many countries is starting to hit hard.

In this context, there is a stark contrast between the People's Republic of China, currently the country with the largest population, and its Asiatic neighbor Japan. While China's universities are turning out about 400,000 engineers every year (and in addition are dramatically increasing their numbers of international students, with a total of nearly 200,000 in 2007, as noted in Finding 3), Japan is facing a dwindling number of young people entering engineering and technology-related fields (Fackler, 2008). The decline is so drastic that industry is going to extremes in advertising and recruiting campaigns to capture engineers or is sending jobs abroad, mainly to Vietnam and India. Part of the decline is due to a dramatically declining population, since Japan has one of the lowest fertility rates in the world, 1.22 in 2008 (CIA, 2008). In addition, "... [A]ccording to educators, executives and young Japanese themselves, the young ... are behaving more like Americans: choosing better-paying fields like finance and medicine, or more purely creative careers, like the arts, rather than following their 'salaryman fathers' into the unglamorous world of manufacturing... The digital technology industry... is already [estimated to be] short half a million engineers" (Fackler, 2008).

Japan is trying to make up for this shortage by recruiting foreigners, but — unlike Americans — the country remains culturally closed, and many foreigners refuse to come. In 1983, Prime Minister Yasuhiro Nakasone presented a plan to raise the annual number of foreign students from 10,428 to 100,000, an accomplishment that took 20 years. For the past few years, however, the number has remained at about 120,000, with 118,498 in 2007 (Ishida, 2008). The Japanese University Consortium for Transnational Education was established in March 2006 with participation by 15 universities. The idea is to send Japanese professors abroad to train future foreign students (up to three years) before they move to Japanese universities as freshmen. This project has worked very well in Malaysia, under a low-interest loan from the Japanese government to finance the program. Now, Prime Minister Yasuo Fukuda wants to raise the number of foreign students to 300,000 (Ishida, 2008). In order to achieve this, the government will designate 30 universities to support foreign students, and there will be more overseas offices to recruit and screen prospective students. Half of the foreign students should be able to find jobs in Japan after graduation.

Many countries in Europe are facing similar population crises. The 2008 collective fertility rate for the European Union is 1.5 (CIA, 2008). Immigrant fertility rates are much higher, but it is still a matter of much debate whether these rates will stay the same or decrease and whether big sectors of the immigrant population can be culturally assimilated. Like Japan, many European universities and institutions have seen the writing on the wall and try to compensate for the diminishing number of national students by actively recruiting international students, creating international degrees, and/or by fostering excellence in their student body through diverse centers of excellence and elite programs, such as those in Japan and Germany.

The United States is sitting on the magical number of 2.1 for its fertility rate (CIA, 2008). Considering that a much smaller percentage of its population joins the Science and Engineering fields than in other countries (see, for instance, Fig. 11.8), keeping up the number of students should be a major concern, as has been pointed out in the previous findings. This is particularly true for SBE&S, since leadership in applications software is no longer determined by access to hardware or middleware, but by highly qualified, long-term supported labor. The efforts of the Technical University of Munich, Japan, and CIMNE in establishing offices and a workforce overseas to recruit exceptional students reflect a new trend and drive up competition. The United States is still the country providing most scholarships and fellowships for international students, and when things get tough, it generally rises to the occasion. Thus, prompted by their declining number of international students, dozens of U.S. educational institutions sent representatives to the recent International Education Exhibition in Shanghai. These efforts, plus easier access to visas for Chinese students combined with devaluation of the U.S. dollar have resulted in an increase in Chinese student enrollment in the United States (*World Journal*, 2008).

The concept of setting up overseas educational recruiting offices can represent an intriguing new possibility for the United States to consider. As an example, consider the case for recruiting Hispanic students. As of mid-2007, the U.S. Hispanic population represented the largest U.S. minority group, accounting for 15.1% of the total population (U.S. Census Bureau, 2007); it is this group that is mainly responsible for keeping the United States at the replacement fertility rate of 2.1. All in all, the United States has been quite positive in fostering the presence of Hispanics in higher education, where their representation is still well below their share of the total population. However, the attitude has been mainly passive: "Since the immigrants [overwhelmingly Mexican] are here, let's work with what we have." Setting up educational recruiting offices or agreements in Latin American countries, on the other hand, would allow the United States to screen for the very best students at a relatively low cost. The "very best" on the other hand, are increasingly being aggressively targeted by European countries. For instance, the "Tour Europosgrados 2008" took place recently in Universidad Nacional de Cuyo, Mendoza, Argentina. Thousands of scientists took part in this "tour," which showcased fellowships, postdoctoral positions, employment opportunities, cooperation programs, etc., sponsored by the European Commission, the German Academic Exchange Service, the Spanish and French Embassies, the British Council, and the agency CampusFrance (Di Bari, 2008). U.S.–Hispanic academic collaboration already exists in some obvious areas, like astronomy and paleontology- and archeology-related areas, but other areas (especially SBE&S!) could definitely benefit from enhanced collaboration. An interesting example in the area of experimental nanotechnology is the 2005 agreement signed by Lucent Technology and the Bariloche Atomic Center in Argentina, where Argentinean scientists are trained by Lucent and the company works on Argentinean projects (Sametband, 2005).

Ultimately, population matters. In this sense both China and India, and now Saudi Arabia, have big winning cards, and even if they are not top competitors right now, chances are that they will be in the not-so-distant future. They are major sources of international students and now are also becoming major emerging destinations for international students in S&E. Countries in Europe and Japan are in a race to get a share of the human capital in the developing world.

11.4 Case Study: The University of Stuttgart — A Success Story

The University of Stuttgart is a research university with a focus on Engineering and the Natural Sciences. Relevant departments in these areas are Civil and Environmental Engineering; Chemistry; Energy Technology, Process Engineering, and Biological Engineering; Computer Science, Electrical Engineering, and Information Technology; Aerospace Engineering and Geodesy; Engineering Design, Production Engineering and Automotive Engineering; and Mathematics and Physics. Key research areas are modeling and simulation, complex systems, communications, materials, technology concepts and assessment, energy and the environment, mobility construction and living, and integrated products and product design.

The university has a number of collaborative programs and an interdepartmental research structure, with the different departments linked through transfer and research centers interacting with national institutions (e.g., Max Planck Institutes, Fraunhofer Institutes, and the German Aerospace Center), international institutions, and industry. The university sits in one of Europe's strongest economic and industrial regions; Bosch, Fischer, Daimler, HP, IBM, Festo, Pilz Deutschland, Porsche, Trumpf, Stihl, Züblin, and BASF all have facilities in the area.

Stuttgart has a number of superlative technical and computing facilities and projects. These include the facilities of the High Performance Computing Center Stuttgart (HLRS) with a computing power of about 220 Tflops; the Gauss Center for Supercomputing (Europe's most powerful high-performance computing alliance among the universities of Jülich, Munich, and Stuttgart); the Visualization Research Center (VISUS), one of the leading facilities in Europe; the Automotive Simulation Centre Stuttgart (ASCS), in close cooperation with industry since 2007; ArchiNeering (Bangkok's new airport is the result of close cooperation between Civil Engineering and Architecture); Baden-Württemberg Astronautics Center (opening in 2010); the International Center for Cultural and Technological Studies (IZKT); SOFIA, a joint U.S.–German project to build a Boeing 747SP equipped with a high-performance mirror telescope; a wind tunnel that performs tests of aerodynamic and aeroacoustic properties of vehicles up to 265 km/hour; and VEGAS, the research facility for subsurface remediation and for the simulation of contamination processes.

The University of Stuttgart also excels in rankings and funding. It ranks number three in Germany in total amount of external funding (Aachen is first), and number one in Germany in external funding per professor (an average of €400,000 per professor). In 2006, it was one of the top grant university recipients in grant ranking by DFG (Deutsche Forschungsgemeinschaft, the German Research Foundation; see the sidebar on the next page on DFG-Provided Support for Collaborative Research Centers and Programs in Germany); in the top three for a range of Engineering study programs as ranked by the Centre for Higher Education Development (CHE) (*Der Spiegel*, Focus 2007); number five in the CHE 2006 research ranking; most successful German University in the 6th EU Framework Program, particularly in the fields of simulation technology, energy, and e-health; and number two in Germany for research visits by international scholarship holders and award-winners of the Humboldt Foundation, 2005.

In order to support its research and educational efforts, the University of Stuttgart is creating new professorships and research positions. These include three new professorial positions on Mathematical Systems Theory, Modeling of Uncertain Systems, and Human–System Interaction and Cognitive Systems. In addition, there are 13 new junior professorships and 7 new post-doctoral positions with up to 2 research associates each, tenure-track options for 4 of the junior professors, and a total of 72 scientific projects.

11.4.1 *The SimTech Excellence Cluster*

The vision of the SimTech Excellence Cluster at the University of Stuttgart, coordinated by Dr.-Ing. Wolfgang Ehlers, is to progress from isolated numerical approaches to integrative systems science. Interestingly, in order to obtain funding, the group gave arguments partly based on the findings of the U.S. NSF Blue Ribbon Panel Report of February 2006: "... [C]hallenges in SBES... involve... multiscale and multiphysics modeling, real-time integration of simulation methods with measurement systems, model validation and verification, handling large data, and visualization. ... [O]ne of those challenges is education of the next generation of engineers and scientists in the theory and practices of SBES" (Oden *et al.*, 2006). DFG leadership has agreed wholeheartedly with this report and provided funding accordingly. The SimTech Excellence Cluster brings €7 million/year for 5 years. This plus other sources of funding allow for a long-term sustained agenda.

DFG-Provided Support for Collaborative Research Centers and Programs in Germany

DFG Collaborative Research Centers (SFB) and Transregional Collaborative Research Centers (TRR): the budget of these centers is about €2–3 million per year for 12 years, in three 4-year terms. Projects include Spatial World Models for Mobile Context-aware Applications, Selective Catalytic Oxidation of C–H Bonds with Molecular Oxygen, Dynamic Simulation of Systems with Large Particle Numbers; Incremental Specification in Context, and Control of Quantum Correlations in

(*Continued*)

Tailored Matter (a transregional project involving the universities of Stuttgart, Tübingen, and Ulm).

DFG Transfer Units (TFB): these units execute transfer from university to industry. Programs include Simulation and Active Control of Hydroacoustics in Flexible Piping Systems, Development of a Regenerative Reactor System for Autothermal Operation of Endothermic High-Temperature Syntheses, Transformability in Multivariant Serial Production, Rapid Prototyping, Computer Aided Modeling and Simulation for Analysis, and Synthesis and Operation in Process Engineering.

DFG Research Units (FOR): these units are smaller than the centers, generally consisting of 4–5 researchers. They include Nondestructive Evaluation of Concrete Structures Using Acoustic and Electromagnetic Echo Methods, Development of Concepts and Methods for the Determination of Reliability of Mechatronics Systems in Early Stages of Development, Noise Generation in Turbulent Flow, Multiscale Methods in Computational Mechanics, Specific Predictive Maintenance of Machine Tools by Automated Condition Monitoring, and Positioning of Single Nanostructures — Single Quantum Devices.

DFG Priority Programs (SPP): these programs bring €1.2–2 million/year, spread over 10–20 projects. Ten new priority programs are opened per year. They include Molecular Modeling and Simulation in Process Engineering, and Nanowires and Nanotubes: From Controlled Synthesis to Function.

Excellence Initiative: the goal of the Excellence Initiative is to foster excellence in science and research and to raise the profile of top performers in the academic and research community by means of three lines of funding: strategies for the future; excellence clusters, and graduate schools. The University of Stuttgart has been successful in two of the Excellence Initiative's three lines of funding, one of which is the Simulation Technology Excellence Cluster (SimTech Excellence Cluster).

Three new structural research elements of types described in the DFG sidebar above have been founded that provide long-term sustainability for the cluster. They are described below. Compared to traditional university departments with their teaching-oriented "vertical" structure, these research centers are "horizontally" oriented, thus comprising researchers and their institutions from various departments under the common roof of a research goal.

(1) Stuttgart Research Centre of Simulation Technology (SRC SimTech): opened on April 1, 2007, this research center is the first one at the university and represents both a scientific research unit and a new structural element acting as a research department with its own organizational and administrational structure, including financial resources (€240,000/year) and personnel.
(2) SimTech Transfer Unit: this unit bundles all activities of the cluster that require uni- or bidirectional communication with external institutions and industry. It will be embedded in the Stuttgart Transfer Centre, whose role is to transfer research results into application, bundle exchange activities with industrial partners, and provide a basis for all future fundraising activities of individual research centers.
(3) Graduate School of Simulation Technology: part of the Stuttgart School of Science and Technology, the Graduate School integrates the activities of the doctoral students supervised by the members of SimTech.

The SimTech Excellence Cluster has created new lines of elite education, which include both BSc and MSc theses in different SimTech research areas. For the MSc, there is flexible study regulation, with more than one supervisor and at least one from abroad. The MSc student is also required to spend one term abroad. The Graduate School of Simulation Technology also offers many other novelties: (1) there is no more separation in departments, it is completely interdisciplinary; (2) it fosters software skills; (3) it has an international exchange program; and (4) there is joint internal/external and international supervision.

In conclusion, the German Government, the University of Stuttgart (and certainly other universities in Germany) are highly aware of the computational challenges at the root of all meaningful innovation in the S&E fields. Accordingly, they have taken decisive action to implement meaningful changes. Thus, as indicated in the sidebar above, new academic structures have been created, such as the DFG–supported Collaborative Research Centers, Transfer Units, Priority Programs (e.g., Molecular Modeling and Simulation in Process Engineering), Excellence Initiatives (e.g., Simulation Technology Excellence Cluster), and new graduate programs (a new educational model from BSc to PhD at the Graduate

School of Simulation Technology). All these programs are strongly supported economically, with several million Euros per year for several years (5–12 years). Both the academics and the government policymakers have understood that a long-term sustained research agenda is the only way this integrative, simulation-based vision can be realized. These initiatives have clearly strengthened the position of the University of Stuttgart as a major global player in simulation technology.

11.5 Conclusions

The conclusions of this chapter are summarized by the findings:

- Increased Asian and especially EU leadership in SBE&S education is due to increased funding and industrial participation. The United States continues to be a leader in most scientific and engineering enterprises, but its lead is decreasing across all S&E indicators: research and development investment, number of scientific publications, number of scientific researchers, etc. For the first time, the number of U.S. doctorates both in the Natural Sciences and in Engineering has been surpassed by those in the European Union and by those in Asia.
- The European Union is investing in new centers and programs for education and training in SBE&S — all of an interdisciplinary nature. New BSc and MSc degrees are being offered in SBE&S through programs that comprise a large number of departments. In many cases, a complete restructuring of the university has taken place in order to create, for instance, MSc degrees, graduate schools (e.g., Stuttgart), or international degrees in simulation technology.
- European and Asian educational/research centers are attracting more international students from all over the world (even from the United States). Special SBE&S programs (in English) are being created for international students. The United States is seeing smaller increases in international enrollment compared to other countries. This is due not only to post-9/11 security measures but also — and increasingly — to active competition from other countries.
- There are pitfalls in interdisciplinary education, including a tradeoff between breadth and depth. In order to solve "grand challenge" problems in a given field, solid knowledge of a core discipline, in addition to computational skills, are absolutely crucial. Another important problem of interdisciplinary research is the evaluation of scientific performance,

which currently is generally carried out within the confines of disciplinary indicators.

- Demand exceeds supply. There is a huge demand in the European Union and Asia for qualified SBE&S students who get hired immediately after their MSc degrees by industry or finance: there is both collaboration and competition between industry and academia.
- There is widespread difficulty in finding students and postdocs qualified in algorithmic and program development. Most countries are competing for leadership in applications software development, where the highest impacts on Science and Engineering are expected to take place. Unfortunately, students and postdocs with the appropriate training in the development of algorithms and programs are hard to recruit, since they are trained primarily to run existing codes rather than to develop the skills necessary for programming.
- Still, the center of applications software development has shifted to the European Union. This is due to well-funded, long-term collaboration efforts; students thrive in these environments. The U.S. system for tenure and promotion, which emphasizes high number and visibility of publications, discourages software development. The merit system should value software contributions as well as publications.
- Population matters. Countries in the European Union and Japan are scampering to recruit foreign students to make up for the dwindling numbers of their youths. This is driving fierce competition for international recruiting. The United States, sitting at exactly the "fertility replacement rate" should pay close attention, since only a small percentage of its population joins the Sciences and Engineering professions. China and India, on the other hand, are "sitting pretty" and are likely to become increasingly strong SBE&S competitors.

References

Central Intelligence Agency (2008). Rank order: total fertility rate, *The World Factbook*, https://www.cia.gov/library/publications/the-world-factbook/rank-order/2127rank.html.

Committee on Science, Engineering, and Public Policy (2007). *Rising Above the Gathering Storm: Energizing and Employing America for a Brighter Economic Future* (National Academies Press, Washington, DC), http://www.nap.edu/catalog.php?record_id=11463#orgs.

Council on Competitiveness. (2007). *Competitiveness Index: Where America Stands* (The Council on Competitiveness, Washington, DC), http://www.compete.org/publications/.

Di Bari, V. (2008). Oferta de becas y posgrados europeos en la UNCuyo, *Los Andes Online*, August 15, http://www.losandes.com.ar/notas/2008/8/15/sociedad-375172.asp.

Fackler, M. (2008). High-tech Japan running out of engineers, *The New York Times*, May 17, http://www.nytimes.com/2008/05/17/business/worldbusiness/17engineers.html.

Ishida, I. (2008). Educational renaissance/preparation key for foreign students, *Daily Yomiuri Online*, July 31, http://www.yomiuri.co.jp/dy/features/language/20080731TDY14001.htm.

Jaschik, S. (2007). The mobile international student, *Inside Higher Ed*, Oct. 10, http://www.insidehighered.com/news/2007/10/10/mobile.

King Abdullah University of Science and Technology (2008). King Abdullah University of Science and Technology and IBM to build one of the world's fastest and most powerful supercomputers, http://www.kaust.edu.sa/news-releases/king-abdullah-university-and-ibm-build-supercomputer.aspx.

Lexington (2008). Help not wanted, *The Economist*, April 10, http://www.economist.com/world/unitedstates/displaystory.cfm?story_id=11016270.

Ministry of Education of the People's Republic of China (2000). *International Students in China*, http://www.moe.edu.cn/english/international_3.htm.

National Science Board (2006). *Science and Engineering Indicators 2006*, vols. 1 and 2 (National Science Foundation, Arlington, VA).

National Science Board (2008). *Science and Engineering Indicators 2008*, vols. 1 and 2 (National Science Foundation, Arlington, VA).

National Science Foundation, Division of Science Resources Statistics (2007). *First-Time, Full-Time Graduate Student Enrollment in Science and Engineering Increases in 2006, Especially Among Foreign Students* (National Science Foundation, Washington, DC), http://www.nsf.gov/statistics/infbrief/nsf08302/.

Oden, J. T., Belytschko, T., Hughes, T. J. R., Johnson, C., Keyes, D., Laub, A., Petzold, L., Srolovitz, D. and Yip, S. (2006). *Revolutionizing Engineering Science through Simulation: A Report of the National Science Foundation Blue Ribbon Panel on Simulation-Based Engineering Science* (National Science Foundation, Washington, DC), http://www.nsf.gov/pubs/reports/sbes_ final_report.pdf.

Organisation for Economic Co-operation and Development (2006). *Main Science and Technology Indicators 2006* (OECD, Publishing, Paris), http://www.oecd.org/document/26/0,3343,en_2649_34451_1901082_1_1_1_1,00.html.

Organisation for Economic Co-operation and Development (2008). *Education Online Database*, http://www.sourceoecd.org/vl=1605451/cl=31/nw=1/rpsv/statistic/s4_about.htm?jnlissn=16081250, See also *Education at a Glance 2007: OECD Indicators* (OECD Publishing, Paris), http://www.oecdbookshop.org/oecd/index.asp.

People's Daily Online staff (2008). Number of international students in China exceeds 190,000 in 2007, *People's Daily Online*, March 14, http://english.peopledaily.com.cn/90001/6373700.html.

Primeur Monthly staff (2008). KAUST announces partnership with UC San Diego to build world's most advanced visualization centre, *Primeur Monthly*, http://enterthegrid.com/primeur/08/articles/monthly/AE-PR-11-08-102.html.

Sametband, R. (2005). Argentina invests US$10 million in nanotechnology, *Sci. Dev. Network*, May 12, http://www.scidev.net/en/news/argentina-invests-us10-million-in-nanotechnology.html.

World Journal staff (2008). Devaluation of U.S. dollar draws international students from China, *World Journal*, March 11.

UNESCO (2007). United Nations Educational, Scientific, and Cultural Organization.

U. S. Census Bureau (2007). U. S. Hispanic population surpasses 45 million; Now 15 percent of total [news release], http://www.census.gov/Press-Release/www/releases/archives/population/011910.html.

Appendix A

BIOGRAPHIES OF PANELISTS AND ADVISORS

A.1 Panelists

A.1.1 *Sharon C. Glotzer (Chair)*

Sharon C. Glotzer is the Stuart W. Churchill Collegiate Professor of Chemical Engineering and Professor of Materials Science and Engineering, at the University of Michigan, Ann Arbor. She also holds faculty appointments in Physics, Applied Physics, and Macromolecular Science and Engineering. She received a B.S. in physics from UCLA in 1987 and a Ph.D. in physics from Boston University in 1993. Prior to Michigan, she worked at the National Institute of Standards and Technology where she held a National Research Council postdoctoral fellowship, and co-founded and directed the Center for Theoretical and Computational Materials Science.

Dr. Glotzer's research focuses on computational nanoscience and simulation of soft matter, self-assembly and materials design, and is sponsored by the Department of Defense, Department of Energy, National Science Foundation, and the J.S. McDonnell Foundation. She has mentored more than 40 PhD and postdoctoral students, has over 140 archival papers, and has given over 220 invited, keynote, and named lectures. Professor Glotzer is a Fellow of the American Physical Society (APS) and holds a prestigious National Security Science and Engineering Faculty Fellowship from the U.S. Department of Defense. She is the recipient of numerous awards, including the Charles M.A. Stine Award from the American Institute of Chemical Engineers (AIChE), the Maria Goeppert-Mayer Award from the American Physical Society, the Presidential Early Career Award for Scientists and Engineers (PECASE), and the Department of Commerce Bronze Medal Award for Superior Federal Service. In 2011 she was elected to the American Academy of Arts and Sciences.

Dr. Glotzer has held elected offices in AIChE and APS, and served on numerous National Academies' committees on such topics as technology surprise, biomolecular materials and processes, modeling, simulation and games, solid state sciences, high performance computing, and defense intelligence. Professor Glotzer serves on numerous editorial and advisory boards, and has provided leadership and input on roadmapping for federal granting agencies on many research topics, including simulation-based engineering and science. She is the co-founding director of a new Virtual School of Computational Science and Engineering. She currently serves as the Director of Research Computing for the University of Michigan College of Engineering, and is the founding director of the new UM Institute for Computational Science & Engineering.

A.1.2 *Sangtae Kim (Vice Chair)*

Sangtae Kim is the inaugural Executive Director of the Morgridge Institute for Research, a new nonprofit medical research institute based in Madison, Wisconsin. Previously, he was the Donald W. Feddersen Distinguished Professor of Mechanical Engineering and Distinguished Professor of Chemical Engineering at Purdue University. From February 2004 to August 2005, he was at the National Science Foundation (on loan from Purdue) to serve as the inaugural director of NSF's Division of Shared Cyberinfrastructure and to help guide NSF's formational investments in cyberinfrastructure.

From 1997 to 2003, Dr. Kim served as Vice president overseeing the R&D IT departments at two pharmaceutical companies, Eli Lilly and Warner Lambert, during the transition to the data-intensive, post-genomic IT environment in the research-based pharmaceutical industry.

From 1983 to 1997, Dr. Kim was a faculty member in the Department of Chemical Engineering at the University of Wisconsin–Madison, progressing from Assistant professor to distinguished Chair professor for his work in mathematical and computational methods for microhydrodynamics (now more commonly known as microfluidics). His computational insights into "hydrodynamic steering" played an influential role in 1994–1995 in the development of fluidic self assembly (FSA), the dominant process used today for manufacture of low-cost RFID (radio frequency identification) tags. In recognition of his teaching and research accomplishments in high performance computing, he was extended a courtesy faculty appointment in the UW-Madison Computer Sciences Department.

Dr. Kim is a member of the National Academy of Engineering. His research citations include the 1993 Allan P. Colburn Award of the

American Institute of Chemical Engineers, the 1992 Award for Initiatives in Research from the National Academy of Sciences and a Presidential Young Investigator award from NSF in 1985. He has an active record of service on science and technology advisory boards of government agencies, the U.S. National Research Council and companies in IT-intensive industries.

A native of Seoul and a product of the "K-11" public schools of Montreal, Dr. Kim received concurrent BSc and MSc degrees (1979) from Caltech and his PhD (1983) from Princeton.

A.1.3 *Peter T. Cummings*

Peter T. Cummings is the John R. Hall professor of chemical engineering at Vanderbilt University. He is also Principal Scientist in the Center for Nanophase Materials Sciences (CNMS) at Oak Ridge National Laboratory, as well as Director of the Nanomaterials Theory Institute within the CNMS. The CNMS is the first of five Department of Energy nanoscience centers, entering full-scale operations on October 1, 2005. Currently, the CNMS has over 200 user projects, with almost 50 of those user projects being in the Nanomaterials Theory Institute.

Dr. Cummings' research interests include statistical mechanics, molecular simulation, computational materials science, computational and theoretical nanoscience, and computational biology. He is the author of over 300 refereed journal publications and the recipient of many awards, including the 1998 Alpha Chi Sigma award (given annually to the member of the American Institute of Chemical Engineers (AIChE) with the most outstanding research contributions over the previous decade) and the 2007 AIChE Nanoscale Science and Engineering Forum award (given to recognize outstanding contributions to the advancement of nanoscale science and engineering in the field of chemical engineering through scholarship, education or service). He is a fellow of the American Physical Society and the American Association for the Advancement of Science (AAAS).

Dr. Cummings received a BMath degree (First Class Honors) at the University of Newcastle (Australia) in 1976 and a PhD in Mathematics at the University of Melbourne (Australia) in 1980.

A.1.4 *Abhijit Deshmukh*

Abhijit Deshmukh is Professor of Industrial and Systems Engineering, and Director of the Institute for Manufacturing Systems at Texas A&M University. Previously, he was Professor of Mechanical and

Industrial Engineering, and Director of the Consortium for Distributed Decision-Making (CDDM) and the Security, Emergency Preparedness and Response Institute (SEPRI) at the University of Massachusetts Amherst. From September 2004 to August 2007, he was a Program Director at the National Science Foundation in the Engineering Directorate and the Office of Cyberinfrastructure. Dr. Deshmukh received his PhD in Industrial Engineering from Purdue University in 1993, and BE degree in Production Engineering from the University of Bombay in 1987.

Dr. Deshmukh was awarded the National Science Foundation Director's Award for Collaborative Integration in 2005, Ralph R. Teetor Educational Award from the Society of Automotive Engineers in 2003, and Milton C. Shaw Outstanding Young Manufacturing Engineer Award from the Society of Manufacturing Engineers in 1999. He was a Lilly Teaching Fellow at the University of Massachusetts from 1999 to 2000.

Dr. Deshmukh's research group is focused on distributed decision-making, with specific interests in complex systems and complexity in decision-making; coordination and inferencing in distributed sensor networks; multiscale decision models; negotiation protocols, contract portfolio selection and dynamic pricing in supply chains; multi-agent models and Cyberinfrastructure for extended enterprises; and life-cycle cost estimation and uncertainty propagation in distributed design.

A.1.5 *Martin Head-Gordon*

Martin Head-Gordon is a Professor of Chemistry at the University of California, Berkeley and is a Faculty Chemist in the Chemical Sciences Division of Lawrence Berkeley National Laboratory working in the area of computational quantum chemistry. He is a member of The International Academy of Quantum Molecular Science.

A native of Australia, Head-Gordon obtained his PhD from Carnegie Mellon under the supervision of John Pople developing a number of useful techniques including the Head-Gordon–Pople scheme for the evaluation of integrals and the orbital rotation picture of orbital optimization. He received the BSc and MSc degrees at Monash University (Australia).

His awards include a 1993 National Science Foundation Young Investigator Award, an Alfred P. Sloan Foundation Research Fellowship (1995–1997), a David and Lucile Packard Fellowship (1995–2000), and the 1998 Medal of the International Academy of Quantum Molecular Sciences. Prior to joining the Berkeley faculty, Dr. Head-Gordon was a postdoctoral researcher at AT&T Bell Laboratories, where he developed unique electronic structure methods for calculating nonadiabatic energy exchange between molecules and surfaces.

At Berkeley, Martin Head-Gordon supervises a group interested in pairing methods, local correlation methods, dual-basis methods, scaled MP2 methods, new efficient algorithms, and very recently corrections to the Kohn-Sham density functional framework. Broadly speaking, wavefunction-based methods are the focus of the research. Dr. Head-Gordon has authored over 200 publications and presented over 150 invited lectures.

Martin is also one of the founders of Q-Chem, Inc. which brings commercial, academic and government scientists worldwide in pharmaceuticals, materials science, biochemistry and other fields a comprehensive *ab initio* quantum chemistry program.

A.1.6 *George Em Karniadakis*

George Em Karniadakis received his SM and PhD degrees from Massachusetts Institute of Technology. He was a Lecturer in the Department of Mechanical Engineering at MIT in 1987, and postdoctoral fellow at the Center for Turbulence Research at Stanford/NASA Ames. He joined Princeton University as Assistant Professor in the Department of Mechanical and Aerospace Engineering with a joint appointment in the Program of Applied and Computational Mathematics. He was a Visiting Professor at Caltech (1993) in the Aeronautics Department. He joined Brown University as Associate Professor of Applied Mathematics and became a full professor in 1996. He has been a Visiting Professor and Senior Lecturer of Ocean/Mechanical Engineering at MIT since September 1, 2000. He was Visiting Professor at Peking University (Fall 2007).

Dr. Karniadakis is a Fellow of the American Physical Society (APS, 2004–), Fellow of the American Society of Mechanical Engineers (ASME, 2003–) and Associate Fellow of the American Institute of Aeronautics and Astronautics (AIAA, 2006–). He is the recipient of the CFD award (2007) by the U.S. Association in Computational Mechanics. His research interests are focused on computational fluid dynamics, stochastic modeling, biophysics, uncertainty quantification and parallel computing. He is the author of three textbooks and more than 250 research papers.

A.1.7 *Linda Petzold*

Linda Petzold is currently Professor in the Department of Computer Science, Professor in the Department of Mechanical Engineering, and

Director of the Computational Science and Engineering Program at the University of California Santa Barbara. She received her PhD in Computer Science from the University of Illinois. She was a member of the Applied Mathematics Group at Sandia National Laboratories in Livermore, California, Group Leader of the Numerical Mathematics Group at Lawrence Livermore National Laboratory, and Professor in the Department of Computer Science at the University of Minnesota.

Dr. Petzold is a member of the U. S. National Academy of Engineering, and a Fellow of the ASME and of the AAAS. She was awarded the Wilkinson Prize for Numerical Software in 1991, the Dahlquist Prize in 1999, and the AWM/SIAM Sonia Kovalevski Prize in 2003. She served as SIAM (Society for Industrial and Applied Mathematics) Vice President at Large from 2000 to 2001, as SIAM Vice President for Publications from 1993 to 1998, and as Editor in Chief of the SIAM Journal on Scientific Computing from 1989 to 1993.

Professor Petzold's research group concentrates on computational science and engineering and systems biology. The computational issues addressed range from development and analysis of numerical methods for problems at scales from stochastic to deterministic, to sensitivity analysis and model reduction, to the design of software environments to make scientific computation more easily accessible.

A.1.8 *Celeste Sagui*

Celeste Sagui, Associate Professor of Physics at North Carolina State University, received her doctorate in physics from the University of Toronto and performed postdoctoral work at McGill University and at the National Institutes of Environmental and Health Sciences (NIEHS). She continues to hold an Intergovernmental Personnel Act (IPA) position in the Laboratory of Structural Biology at NIEHS.

Dr. Sagui is exploring properties of condensed matter systems and biomolecules, focusing on methodological developments for the accurate and efficient treatment of electrostatics in large-scale biomolecular simulations, studies of DNA structure and other selected biomolecules, molecular and ion solvation, ordering in modulated condensed matter systems for nanotechnological applications, and organic molecules on surfaces. To explore the interesting properties of these systems, she uses a range of computational methods such as quantum chemistry, density functional theory, classical molecular dynamics and phase field models and hydrodynamical equations.

Dr. Sagui has received many awards, including the NSF CAREER Development Award; NSF POWRE Award; SLOAN Postdoctoral Fellowship in Computational Biology; Ontario Graduate Scholarship (OGS), Toronto; Provost Seeley Fellowship, Trinity College, University of Toronto; International Student Differential Fee Waiver Scholarship, Toronto; Connaught Scholarship, Toronto; and the "Diploma de Honor" awarded by the University of San Luis, Argentina for the highest marks in the graduating class for Physics and Mathematics.

A.1.9 *Masanobu Shinozuka*

Masanobu Shinozuka, a Member of the National Academy of Engineering since 1978, is currently Distinguished Professor and Chair of the Department of Civil and Environmental Engineering at the University of California Irvine. He is also Norman Sollenberger Professor of Civil Engineering Emeritus at Princeton University. He received his BS and MS degrees from Kyoto University (Japan) and his PhD degree in 1960 from Columbia University.

Professor Shinozuka's research interests include continuum mechanics, stochastic processes, structural dynamics and control, earthquake engineering, reliability of infrastructure and lifeline systems.

Prior to joining the University of California Irvine, he taught at the University of Southern California, Princeton University, the State University of New York at Buffalo where he also served as Director of the NSF National Center for Earthquake Engineering Research (on leave from Princeton University), and Columbia University.

Professor Shinozuka is an Honorary member of ASCE, Fellow of ASME, Senior Member of AIAA, Editor-Emeritus of *The Journal of Probabilistic Engineering Mechanics*, and present and past member of the editorial boards of many reputable technical journals. He is the recipient of various national and international awards and recognition such as ASCE's major research awards, including the Freudenthal, Newmark, von Karman and Scanlan medals, and has been elected to the Russian Academy of Architecture and Construction Science.

Professor Shinozuka continues to be active, currently serving as PI for a $5.5 M joint venture under NIST's Technology Innovation Program entitled, "Next Generation SCADA for Protection and Mitigation of Water System Infrastructure Disaster."

A.2 Advisors

A.2.1 *Tomás Díaz de la Rubia*

Tomás Díaz de la Rubia joined LLNL as a postdoc in 1989 after completing his PhD in physics at the State University of New York at Albany. He carried out his thesis research in the Materials Science Division at Argonne National Laboratory and in the Materials Science Department at University of Illinois at Urbana-Champaign. The focus of his scientific work has been the investigation, via large-scale computer simulation, of defects, diffusion, and microstructure evolution in extreme environments.

At LLNL, he first worked on materials issues for the fusion program and then joined the Chemistry and Materials Sciences (CMS) Directorate in 1994. Between 1994 and 1996, he focused his research activities around the development of physics-based predictive models of ion implantation and thin film growth for semiconductor processing in collaboration with Bell Labs, Intel, Applied Materials, IBM and other semiconductor corporations. Between 1994 and 2002, he was also involved in the development of multiscale models of materials strength and aging in irradiation environments and worked in the Advanced Simulation and Computing Program developing models of materials strength.

In 1999, he became group leader for Computational Materials Science and helped build and lead an international recognized effort in computational materials science at LLNL. Between 2000 and 2002, he served as the CMS Materials Program Leader for the National Ignition Facility (NIF), where he focused on optical materials and target development for NIF applications.

Dr. Tomás Díaz de la Rubia was selected as the Associate Director for Chemistry and Materials Science in 2002. Currently, he leads the Chemistry, Materials, Earth, and Life Sciences Directorate, formed in 2007 after merging CMS and the previous Biosciences and Energy and Environment directorates.

Dr. Tomás Díaz de la Rubia has published more than 140 peer reviewed articles in the scientific literature, has chaired numerous international conferences and workshops, and has edited several conference proceedings and special journal issues. He belongs to the editorial board of five major scientific journals, and continues to serve in numerous national and international panels. He was elected a fellow of the American Physical Society in 2002, and is currently the Vice chair (chair elect) of the Division of Computational Physics. He is a fellow of the American Association for the

Advancement of Science (AAAS), and he served as an elected member of the board of directors of the Materials Research Society between 2002 and 2005.

A.2.2 *Jack Dongarra*

Jack Dongarra received a Bachelor of Science in Mathematics from Chicago State University in 1972 and a Master of Science in Computer Science from the Illinois Institute of Technology in 1973. He received his PhD in Applied Mathematics from the University of New Mexico in 1980. He worked at the Argonne National Laboratory until 1989, becoming a senior scientist. He now holds an appointment as University Distinguished Professor of Computer Science in the Computer Science Department at the University of Tennessee and holds the title of Distinguished Research Staff in the Computer Science and Mathematics Division at Oak Ridge National Laboratory (ORNL), Turing Fellow at Manchester University, and an Adjunct Professor in the Computer Science Department at Rice University. He is the Director of the Innovative Computing Laboratory at the University of Tennessee. He is also the Director of the Center for Information Technology Research at the University of Tennessee which coordinates and facilitates IT research efforts at the University.

He specializes in numerical algorithms in linear algebra, parallel computing, the use of advanced-computer architectures, programming methodology, and tools for parallel computers. His research includes the development, testing and documentation of high quality mathematical software. He has contributed to the design and implementation of the following open source software packages and systems: EISPACK, LINPACK, the BLAS, LAPACK, ScaLAPACK, Netlib, PVM, MPI, NetSolve, Top500, ATLAS, and PAPI. He has published approximately 200 articles, papers, reports and technical memoranda and he is coauthor of several books. He was awarded the IEEE Sid Fernbach Award in 2004 for his contributions in the application of high performance computers using innovative approaches and in 2008 he was the recipient of the first IEEE Medal of Excellence in Scalable Computing. He is a Fellow of the AAAS, ACM, and the IEEE and a member of the National Academy of Engineering.

A.2.3 *James Johnson Duderstadt*

James J. Duderstadt is President Emeritus and University Professor of Science and Engineering at the University of Michigan.

Dr. Duderstadt received his baccalaureate degree in electrical engineering with highest honors from Yale University in 1964 and his doctorate in Engineering Science and Physics from the California Institute of Technology in 1967. After a year as an Atomic Energy Commission Postdoctoral Fellow at Caltech, he joined the faculty of the University of Michigan in 1968 in the Department of Nuclear Engineering. Dr. Duderstadt became Dean of the College of Engineering in 1981 and Provost and Vice President for Academic Affairs in 1986. He was appointed as President of the University of Michigan in 1988, and served in this role until July 1996. He currently holds a university-wide faculty appointment as University Professor of Science and Engineering, directing the University's program in Science, Technology, and Public Policy, and chairing the Michigan Energy Research Council coordinating energy research on the campus.

Dr. Duderstadt's teaching and research interests have spanned a wide range of subjects in science, mathematics, and engineering, including work in areas such as nuclear fission reactors, thermonuclear fusion, high powered lasers, computer simulation, information technology, and policy development in areas such as energy, education, and science.

During his career, Dr. Duderstadt has received numerous national awards for his research, teaching, and service activities, including the E. O. Lawrence Award for excellence in nuclear research, the Arthur Holly Compton Prize for outstanding teaching, the Reginald Wilson Award for national leadership in achieving diversity, and the National Medal of Technology for exemplary service to the nation. He has been elected to numerous honorific societies including the National Academy of Engineering, the American Academy of Arts and Science, Phi Beta Kappa, and Tau Beta Pi.

Dr. Duderstadt has served on and/or chaired numerous public and private boards. These include the National Science Board; the Executive Council of the National Academy of Engineering; the Committee on Science, Engineering, and Public Policy of the National Academy of Sciences; the Nuclear Energy Research Advisory Committee of the Department of Energy; the Big Ten Athletic Conference; the University of Michigan Hospitals; Unisys; and CMS Energy.

He currently serves on or chairs several major national study commissions in areas including federal science policy, higher education, information technology, and energy sciences, including NSF's Advisory Committee on Cyberinfrastructure, the National Commission on the Future of Higher Education, the AGB Task Force on the State of the University Presidency, the Intelligence Science Board, and the Executive Board of the AAAS.

A.2.4 *J. Tinsley Oden*

J. Tinsley Oden was the founding Director of the Institute for Computational Engineering and Sciences (ICES), which was created in January of 2003 as an expansion of the Texas Institute for Computational and Applied Mathematics, also directed by Oden for over a decade. The institute supports broad interdisciplinary research and academic programs in computational engineering and sciences, involving 4 colleges and 17 academic departments within UT Austin.

An author of over 500 scientific publications: books, book chapters, conference papers, and monographs, Dr. Oden is an editor of the series Finite Elements in Flow Problems and of Computational Methods in Nonlinear Mechanics. Among the 50 books he has authored or edited are Contact Problems in Elasticity, a six-volume series: Finite Elements, An Introduction to the Mathematical Theory of Finite Elements, and several textbooks, including Applied Functional Analysis and Mechanics of Elastic Structures, and, more recently, A Posteriori Error Estimation in Finite Element Analysis, with M. Ainsworth. His treatise, Finite Elements of Nonlinear Continua, published in 1972 and subsequently translated into Russian, Chinese, and Japanese, is cited as having not only demonstrated the great potential of computational methods for producing quantitative realizations of the most complex theories of physical behavior of materials and mechanical systems, but also established computational mechanics as a new intellectually rich discipline that was built upon deep concepts in mathematics, computer sciences, physics, and mechanics. Computational Mechanics has since become a fundamentally important discipline throughout the world, taught in every major university, and the subject of continued research and intellectual activity. Oden has published extensively in this field and in related areas over the last three decades.

Dr. Oden is an Honorary Member of the American Society of Mechanical Engineers and is a Fellow of six international scientific/technical societies: IACM, AAM, ASME, ASCE, SES, and BMIA. He is a Fellow, founding member, and first President of the U.S. Association for Computational Mechanics and the International Association for Computational Mechanics. He is a Fellow and past President of both the American Academy of Mechanics and the Society of Engineering Science. Among the numerous awards he has received for his work, Dr. Oden was awarded the A. C. Eringen Medal, the Worcester Reed Warner Medal, the Lohmann Medal, the Theodore von Karman Medal, the John von Neumann medal, the Newton/Gauss Congress Medal, and the Stephan P. Timoshenko Medal. He was also knighted as "Chevalier des Palmes Academiques" by the French government and he holds five honorary doctorates, Honoris Causa, from universities in Portugal (Technical University of Lisbon), Belgium (Faculte Polytechnique), Poland (Cracow University of Technology), the United States (Presidential Citation, The University of Texas at Austin), and France (Ecole Normale Superieure Cachan (ENSC)).

In 2004, Dr. Oden was among the seven UT-Austin engineering faculty listed as the most highly cited researchers in the world from 1981 to 1999 in refereed, peer-reviewed journals, according to the International Scientific Index.

Dr. Oden is a member of the U.S. National Academy of Engineering and the National Academies of Engineering of Mexico and of Brazil. He serves as Co-Chairman of the Accelerated Strategic Computing Initiative (ASCI) Panel for Sandia National Laboratories. He is a Member of the IUTAM Working Party 5 on Computational Mechanics and serves on numerous organizational, scientific and advisory committees for international conferences and symposiums. He is an Editor of Computer Methods in Applied Mechanics and Engineering and serves on the editorial board of 27 scientific journals.

Dr. Oden has worked extensively on the mathematical theory and implementation of numerical methods applied to problems in solid and fluid mechanics and, particularly, nonlinear continuum mechanics. His current research focuses on the subject of multiscale modeling and on new theories and methods his group has developed for what they refer to as "adaptive modeling." The core of any computer simulation is the mathematical model used to study the physical system of interest. They have developed methods that estimate modeling error and adapt the choices of models to control error. This has proven to be a powerful approach for multiscale problems. Applications include semiconductors manufacturing at the nanoscale level. Dr. Oden, along with ICES researchers, is also working on adaptive control methods in laser treatment of cancer, particularly prostate cancer. This work involves the use of dynamic-data-driven systems to predict and control the outcome of laser treatments using our adaptive modeling strategies.

A.2.5 *Gilbert S. Omenn*

Gilbert S. Omenn is currently Director, Center for Computational Medicine and Biology at the University of Michigan. He served as Executive

Vice President for Medical Affairs and as Chief Executive Officer of the University of Michigan Health System from 1997 to 2002. He was formerly Dean of the School of Public Health and Professor of Medicine and Environmental Health, University of Washington, Seattle. His research interests include cancer proteomics, chemoprevention of cancers, public health genetics, science-based risk analysis, and health policy. He was principal investigator of the beta-Carotene and Retinol Efficacy Trial (CARET) of preventive agents against lung cancer and heart disease; Director of the Center for Health Promotion in Older Adults; and Creator of a university-wide initiative on Public Health Genetics in Ethical, Legal, and Policy Context while at the University of Washington and Fred Hutchinson Cancer Research Center. He served as Associate Director, Office of Science and Technology Policy, and Associate Director, Office of Management and Budget, in the Executive Office of the President in the Carter Administration. He is a longtime Director of Amgen Inc. and of Rohm & Haas Company. He is a member of the Council and leader of the Plasma Proteome Project for the international Human Proteome Organization (HUPO). In 2004, he became president-elect of the American Association for the Advancement of Science (AAAS).

Dr. Omenn is the author of 449 research papers and scientific reviews and author/editor of 18 books. He is a member of the Institute of Medicine of the National Academy of Sciences, the American Academy of Arts and Sciences, the Association of American Physicians, and the American College of Physicians. He chaired the presidential/congressional Commission on Risk Assessment and Risk Management ("Omenn Commission"), served on the National Commission on the Environment, and chaired the NAS/NRC/IOM Committee on Science, Engineering and Public Policy.

He is active in cultural and educational organizations, and is a musician and tennis player. Omenn received his BA from Princeton, the MD, magna cum laude, from Harvard Medical School, and a PhD in genetics from the University of Washington.

A.2.6 *David E. Shaw*

David E. Shaw serves as Chief Scientist of D. E. Shaw Research and as a Senior Research Fellow at the Center for Computational Biology and Bioinformatics at Columbia University. He received his PhD from Stanford University in 1980, served on the faculty of the Computer Science Department at Columbia until 1986, and founded the D. E. Shaw group in 1988. Since 2001, Dr. Shaw has devoted his time to hands-on research in the field of computational biochemistry. He is now personally involved in the development of new algorithms and machine architectures for high-speed molecular dynamics simulations of biological macromolecules, and in the application of such simulations to basic scientific research in structural biology and biochemistry and to the process of computer-aided drug design. Although he leads the lab's research efforts in his role as Chief Scientist, his focus is largely technical, with limited involvement in operational and administrative management.

In 1994, President Clinton appointed Dr. Shaw to the President's Council of Advisors on Science and Technology. He is a fellow of the American Association for the Advancement of Science, and was elected to its Board of Directors in 1998. Dr. Shaw is also a fellow of the American Academy of Arts and Sciences, and serves on the Computer Science and Telecommunications Board of the National Academies.

A.2.7 *Martin Wortman*

Martin Wortman's current research activities focus on technology assessment. His work is concerned with the development of computationally based predictive models that characterize uncertainty associated with the value of risk encumbered technologies; this work appeals to tenets of optimal gambling. In this context, assessment treats the development, design, deployment, and operation of technology as high-stakes wagers having benefits and liabilities that cannot be predicted with certainty. Often, these wagers must be executed when abundant empirical analyses are not available and uncertainty cannot be completely characterized. Key research issues arise when a unique characterization of the probability law on value (i.e., risk) is unavailable (as is often the case with one-and-off-bets). Wortman develops and explores new computational methods, supporting evaluation, assessment, and wagering that are at the interface of probability, stochastic processes, and high-performance computing.

Dr. Wortman teaches advanced course work in Probability, Stochastic Processes, Operations, Reliability & Risk, and Technology Assessment. He teaches introductory graduate courses in Operations Research Methods and Computational Methods. He teaches undergraduate courses in Operations Research Modeling, System Operations, Production Systems, and System Simulation.

Dr. Wortman maintains an active service role in several professional societies. In addition to organizing conference activities and workshops, he has served as Associate Editor for *Operations Research*, Department Editor of Applied Probability and Engineering Statistics for the *IIE Transactions*, Department Editor of Manufacturing for the *IIE Transactions*, and he is past Editor-in-Chief of the *IEEE Transactions on Reliability*.

Appendix B

SURVEY QUESTIONNAIRE

The following questions were developed by the U.S. delegation to prepare for the visit. We hope that they will help you to understand some of our objectives during this tour of leading European sites involved in simulation-based engineering and science (SBE&S) research. We do not expect detailed answers to each of these questions. Therefore, please feel free to examine the list and determine which questions would be most appropriate for you and your organization. We would like to emphasize that we are not seeking proprietary information.

It is our goal that discussion of issues addressed by these questions will lead to a productive exchange of views that will benefit programs in SBE&S research in both of our countries. Of course, our delegation is prepared to share its perspectives on research activities in the US.

Our study has three primary areas for which we are assessing the current state and the future trends of SBE&S research. The primary thematic areas are:

- Materials
- Energy and sustainability, and
- Life sciences and medicine

We are also interested in the impact of infrastructure and other crosscutting issues on the advancement of the field of SBE&S. We have thus identified eight issues applicable to all of the thematic areas. These are:

- Multiscale simulation
- Validation, verification, and quantifying uncertainty
- Simulation software
- Big data and visualization

- Engineering design
- Next-generation algorithms and high performance computing
- Education and training
- Funding, organization, and collaboration

B.1 General

What are the major needs, opportunities or directions in SBE&S research over the next 10- and 20-year time frames?

What are the national and/or regional funding opportunities that would support research to meet these needs, and/or take advantage of these opportunities? Are these funding opportunities expanding?

B.2 Materials/Energy and Sustainability/Life Sciences and Medicine

What major breakthroughs in these fields will require SBE&S; and which are you and or your colleagues pursuing? Within your institution, region, or country, are there identified targets of opportunity for applications of simulation either for scientific research or for engineering applications in these fields?

Which problems could benefit most from a 1–2 order-of-magnitude increase in computational power?

What are examples of major SBE&S successes or failures in these fields?

Do investigators, laboratories and institutions receive any financial compensation for patented inventions derived from their simulations?

Have any start-up companies spun-off based on simulation efforts in your lab? If so please describe them.

B.3 Multiscale Simulation

Describe efforts and advances within your institution to couple multiple simulation methods in order to bridge multiple length and/or time scales.

Is the development of integrated multiscale modeling environments a priority in research funding in your country or region?

B.4 Validation, Verification, and Quantifying Uncertainty

Describe efforts and advances within your institution to validate and verify codes and to quantify uncertainty in simulation-based predictions?

B.5 Simulation Software

What percentage of code used in your group is developed in-house? What percentage of code is commercial? What percentage is open source? What percentage has been developed by others (e.g., under contract or by acquisition)? What are the biggest issues in using models/simulations developed by others? How easy/difficult is it to link codes to create a larger or multifaceted simulation environment?

Who owns the intellectual property rights (IP) to the codes developed in your group?

How do you deal with liability issues for products developed with codes from other sources?

B.6 Big Data and Visualization

What type of data and visualization resources do you need (and have access to) for SBE&S research?

B.7 Engineering Design

What type of models/codes do you use, develop or conduct basic research on pertaining to different phases of engineered system design (conceptual, parametric optimization, operational/control)?

What are the data requirements for a comprehensive life-cycle model of an engineered system? Do you have repositories in place that provide the data?

How do you couple output of the models/simulations to the decision making process (including quantification of uncertainty/error in the predictions)?

What is the curriculum for training doctoral students in all aspects of designing engineered systems using simulation/modeling tools?

Are there efforts within your institution to couple physics-based models with macroscopic logic or econometric models?

B.8 Next-Generation Algorithms and High Performance Computing

Would you characterize the SBE&S research in your lab as needing and/or using primarily desktop, teraflop, or petaflop computing resources?

What are the computational bottlenecks in your simulation problems? What solutions exist now and in the near future, either from a hardware perspective, algorithm perspective, or both?

What is the state of the art in your community in discrete event simulations (model size, parallelization or thread limits, special techniques — time warp, rollback, etc.)?

B.9 Education and Training

Is there a formal scientific computing or computational science and engineering graduate program at your institution? Or, if you are in a company, are their educational institutions/programs in your country or region that prepare students in formal scientific computing or computational science and graduate engineering effectively?

What level of computing expertise do your incoming graduate students or employees possess? Do you feel that they have the necessary background to conduct research in SBE&S or do they need extensive preparation after arriving before they can be productive researchers?

What kind of training is available in your organization for simulation and high performance computing? Is it adequate? Do you believe it will be adequate to address computing on multi-core and petascale architectures? What plans are in place for training programs in next-generation computing?

What fraction of students in your institution study/work in the area of SBE&S, and how has this fraction changed over the past 5–7 years?

What fraction of graduate students/postdocs in SBE&S comes from abroad? From which country do these researchers originate? How many students in your country would you estimate go abroad to earn PhDs in SBE&S?

After completing a PhD in an SBE&S-related field, what route do your students take? Do they accept postdoctoral positions in your country or abroad? What fraction eventually obtains permanent jobs in SBE&S? Is this a desired career path for students? Do they find many job opportunities related to their training in SBE&S? How does industry view students with training in SBE&S?

B.10 Funding, Organization, and Collaboration

What are the roles of academic, government and industrial laboratories in SBE&S research in your country?

Who pays for, and/or is responsible for, the development and sustainability of SBE&S infrastructure (include long term data storage costs, code maintenance, open-source software, etc.)?

What is the funding situation for SBE&S in your country? Example considerations: what are the major sources of funding? Over the past 5 years, has funding of SBE&S increased, decreased or remained roughly constant relative to funding of all of science and engineering research? Is most SBE&S research funded for single investigators, small teams, or large teams? What is the typical duration of SBE&S funding?

Appendix C

BIBLIOMETRIC ANALYSIS OF SIMULATION RESEARCH

Grant Lewison
Evaluametrics Ltd, 50 Marksbury Avenue, Kew, Richmond, Surrey, TW9 4JF, UK

C.1 Introduction

This analysis was carried out in order to assist the panel in two ways:

- to compare the outputs of the USA and other leading countries in simulation research, in terms of both volume and impact, so as to evaluate the position of the USA relative to that of other countries and groups of nations, and establish time trends; and
- to identify the institutions in Europe and the Far East that were active in simulation research, so as to inform the panel's selection of places to visit.

Bibliometrics is the quantitative study of publications, normally of papers in the peer-reviewed serial literature. It can provide objective evaluation data as a complement to the views of an expert panel, and is increasingly being used in this way on a national and institutional level, even sometimes on an individual level. However the conclusions drawn from bibliometrics analysis depend on the accuracy with which the subject area of interest (here, simulation research) is defined, and hence on the database of the bibliographic details of papers that has been created for analysis.

In this study, the original data were drawn from the Science Citation Index (SCI) on CD-ROM for the 10 years 1996–2005; some additional data were taken from the Web of Science (WoS) version (which has a somewhat wider journal coverage) for more recent years to allow for any later changes to be seen. All the data were based on articles and reviews only because they alone embody substantive research findings.

For the first task, the comparison of the USA with other countries, data were obtained on both numbers of papers and also some measures of esteem or impact — percentage of reviews, potential citation impact (based on journal citation impact factors) and actual citation impact (citations to the individual papers). These are useful partial indicators of merit, and if they agree, one can have more confidence that the message they are conveying is reliable. For the second task, the identification of possible sites for the panel to visit, listings were made of the numbers of papers from different locations and an indication was given on what type of simulation was of primary interest (biological, chemical or physical).

C.2 Methodology

The first step was to write down a short definition of the subject area of simulation research. This was done by Professor Sharon Glotzer, the panel chairman, at a meeting with the author in Ann Arbor, MI, in April 2007, and it reads as follows:

> Simulation involves the application of mathematical models using a computer to the study of the underlying physical and chemical processes, and prediction of the behavior and properties of systems, including natural and artificial materials, flow in liquids and gases, energy at all scales including the cellular level, and biomedical sequelae.

This definition states both what is included and what is excluded, and served to define a "filter" that, when applied to the SCI would selectively identify relevant papers. The filter was developed by an interactive process between Professor Glotzer and the author. It consisted of three parts: specialist journals (Table C.1), specialist title words (Table C.2) and negative title words (Table C.3). Papers were selected if they were either in one of the named journals, or had one or more of the title words, or both, but were de-selected if they also contained one or more of the negative title words. The latter were mainly designed to exclude not only papers involving laboratory animals, but also experiments unless they were specifically designed to check simulations.

The filter was used to generate samples of papers that could be marked for relevance. This was done subsequently by three members of the panel,

Table C.1. List of specialist simulation journals (29-character max. abbreviated names).

Calphad[a]	*J Comput Chem*
Combust Theory Model	*J Comput Neurosci*
Comput Biol Chem	*J Comput Phys*
Computation Mech	*Macromol Theory Simul*
Comput Method Appl Mech Eng	*Math Model Method Appl Sci*
Comput Phys Commun	*Med Biol Eng Comput*
Comput Fluids	*Model Simul Mater Sci Eng*
Int J Numer M[a]	*Mol Simulat*
J Comput Acoust	*Theor Comput Fluid Dynamics*
J Comput Appl Math	

[a]Any character(s) or none.

Table C.2. List of positive title words used to define the simulation filter.

1st Principles	Dynamic[a] and model[a]	Monte Carlo
ab initio	Finite difference	Multiscale and model[a]
Bayesian	Finite element[a]	Neural net[a]
Boundary element[a]	Finite volume	Numeric[a] and model[a]
Cellular automata	Integral equation[a]	Numerical calculation[a]
CFD	Kinetic[a] and model[a]	Numerical prediction
Comput[a] and model[a]	Lattice Boltzmann	Particle in cell
Computational	Lattice gas	Phase diagram[a]
Continuous[a] optimiz[a]	Lennard Jones	Phase equilibri[a]
Density functional	Markov chain[a]	Simulat[a]
Discrete element[a]	Mathemat[a] and model[a]	Theoretical and computational
Dissipative particle dynamics	MCMC	
Dynamic optimization	Molecular dynamics	

[a]Any character(s) or none.

Table C.3. List of negative title words used to disqualify a paper for the simulation file.

Experiment[a]	Not simulat[a]	Pig	Rat
Mice	Mouse	Pigs	Rats
Monkey[a]	Murine	Rabbit[a]	

[a]Any character(s) or none.

and the precision, p, was 0.86 while the recall, r, was only 0.45. These figures suggest that the filter could have been further improved by the addition of extra title words, or additional journals, in order to increase its recall.

Table C.4. Major fields used to classify the downloaded simulation papers, and % in each.

Field	%	Field	%	Field	%
Physics	29.6	Mathematics	8.0	Clin. medicine	5.7
Engineering and techn	21.5	Earth and space	7.9	Biology	3.2
Chemistry	16.2	Biomed. research	7.1	Other	0.8

The filter was applied to the SCI on CD-ROM, and all papers (articles and reviews) from the 10 years, 1996–2005, had their bibliographic details (authors, title, document type, full source and addresses) downloaded to three MS Excel files, covering the years 1996–1999, 2000–2002 and 2003–2005. Altogether, there were just over 149,000 papers in the three files.

The papers were now classified by major field, on the basis of the journals in which they were published, using a scheme devised by CHI Research Inc. (now The Patent Board, a subsidiary of IPIQ Inc.) for the US National Science Foundation. Nine fields were used (there are some others with small numbers of papers included in "others"), as listed in Table C.4, which also gives the percentage of papers in each.

The percentage of reviews is a new measure of research esteem and gives a simple indication of how a country's (or an institution's) senior researchers are seen by journal editors. It was calculated for papers from different countries. These percentages can be compared with the corresponding values for all science papers.

The potential citation impact of each paper was determined from the five-year mean citation score of papers in the same journal and year. This is also a simple indicator of research esteem, and although it has been criticized as being inferior to counts of citations to actual papers, it actually measures something different. Values for the year 2002 for some leading journals are shown in Table C.5.

The actual citation impact of the simulation papers was determined from another version of the SCI, the Web of Science (WoS). This has a rather wider journal coverage than the CD-ROMs, and this affects national outputs differently — generally boosting those from Far Eastern countries such as China more than those from European and North American ones. The simulation filter was first applied to the WoS in order to learn about any recent changes in national outputs (with an overlap to earlier years so that outputs could be calibrated against those from the CD-ROMs). The WoS also provided citation data on individual papers and groups (provided that they

Table C.5. Mean five-year citation score (i.e., cites in 2002–2006 to papers published in 2002) for papers in some journals much used by simulation researchers.

Journal	PCI
Physical Review Letters	24.1
Journal of Physical Chemistry B	15.4
Journal of Chemical Physics	12.3
Journal of Computational Physics	9.3
International Journal for Numerical Methods in Engineering	6.6
International Journal for Numerical Methods in Fluids	3.6
Journal of Computational and Applied Mathematics	2.6

numbered fewer than 10,000). This allowed average actual citation scores for simulation papers from different countries to be determined, but only on an integer count basis. (A paper with one US address and two from France would count unity for each on integer counting, and 0.33 and 0.67 respectively on fractional counting.) However, the determination of potential citation impact was also done on a fractional count basis, which is technically more accurate because it allocates credit on a proportional basis.

Since one of the main purposes of the bibliometric analysis was to determine the relative position of the USA in simulation research, comparisons were made with selected countries in Europe and Asia, as listed in Table C.6. The digraph ISO codes are used hereinafter to designate the different countries: they are used as the last element in web domain names and so have become reasonably familiar to web users.

The table lists 8 of 12 western European countries that have been taken as a group for comparison with the USA, and five Asian countries that may be of interest to the panel.

Table C.6. List of countries used for the bibliometric analysis.

Code	Country	Code	Country	Code	Country
AU	Australia	EUR12	(see note)	PL	Poland
BR	Brazil	FR	France[a]	RU	Russia
CA	Canada	IN	India	SE	Sweden[a]
CH	Switzerland[a]	IT	Italy[a]	TW	Taiwan
CN	China (Peoples Rep)	JP	Japan	UK	United Kingdom[a]
DE	Germany[a]	KR	South Korea	US	United States
ES	Spain[a]	NL	Netherlands[a]		

[a] Any character(s) or none.

C.3 Results: National Comparisons

Figure C.1 shows the numbers of simulation papers, year by year, with for comparison the number of papers in the SCI, both from the CD-ROMs. Because of late processing, the tally of 2005 simulation research papers is expected to be about 10% low, so an allowance has been made for this. Data for 2006 and 2007 have been taken from the WoS. However in 2004 and 2005, there were more papers in this version of the SCI, as shown in Table C.7. The data suggest that the coverage of simulation research was about 41% greater in these years, and so the WoS outputs have been

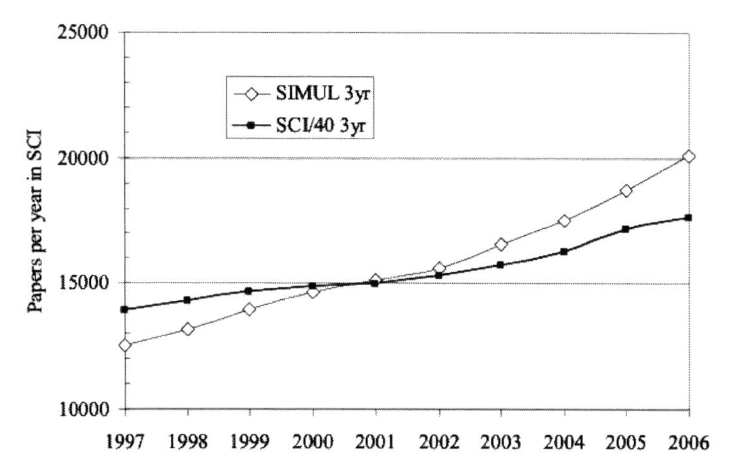

Fig. C.1. Number of papers per year (three-year running means) in simulation research (open diamonds, light line) and in all science (/40; solid squares, heavy line).

Table C.7. Comparison of WoS and CD-ROM outputs of simulation research and all SCI papers for 2004 and 2005.

	WoS		CD-ROM		Ratio	
Year	SCI	SIMUL	SCI	SIMUL	SCI	SIMUL
2004	861,651	25,138	621,998	17,936	1.39	1.40
2005	913,393	27,742	703,805	19,563	1.30	1.42
2006	955,610	30,112	*707,859*	*21,356*		
2007	941,543	27,926	*697,439*	*19,806*		

Figures in italics are estimated on the basis of the mean ratio of WoS to CD-ROMs.

Note: CD-ROM data for 2005 increased by 10% to allow for late processing.

reduced by this amount to give estimated values for the CD-ROMs (which were not available). Overall outputs were only about 34% higher.

It is clear that simulation research is growing much faster than science overall, the annual average percentage growth rate being 5.0% compared with 2.5% for all science.

The outputs of papers from individual countries are shown as fractional counts in Table C.8 for the three periods, 1996–1999, 2000–2002 and 2003–2005. Data are also given for the 12 European countries as a group (with fractional counts the numbers can be simply added) and for the five "Asian tigers" of China, India, Singapore, South Korea and Taiwan (AS5).

Both the USA and Europe12 have reduced their share of world papers (by 12% and 8% of their 1996–1999 presences in 2003–2005), but their absolute numbers have increased in successive periods. The increase in share has gone to the Asian countries, particularly China, whose output increased from 308 papers in 1996 to 1691 in 2005 (after correction for late processing). Its output overtook that of Japan in 2004, and has been growing at a massive 18.5% per year, compared with 3.7% for Europe12 and only 2.8% for the USA. The other Asian countries' outputs are growing at intermediate rates: 4.7% for Japan but 11.5% for South Korea.

It is worthwhile to see how these outputs compare with the countries' overall production of scientific papers: the ratio of percentage presences is their relative commitment (RC) to simulation research. This is shown in chart form for the whole decade in Fig. C.2, which demonstrates that China has much the largest RC (among leading nations), and that the USA (and Canada) do slightly less simulation research than their overall presence in science would suggest. Japan appears to be relatively inactive in this field. Over the decade, Switzerland and Sweden have increased their RC, but South Korea, China and Russia, and in Europe, Germany and the UK, have decreased it.

The percentage of reviews is a simple measure of the esteem in which a country's senior researchers are held, but it needs to be normalized with respect to the average value for all world papers to give a ratio. Within the total numbers of papers, the percentage of reviews has been steadily growing, as shown in Fig. C.3. However, it is clear that simulation has a much smaller percentage of reviews than the mean for all science, and that it has not been growing. This has to be taken into account when the performance of individual countries is considered.

Figure C.4 shows this ratio both for simulation (gray bars) and for all science (white bars) as the average ratios to the world values for the

Table C.8. Annual outputs of simulation papers (fractional counts) from leading countries and regions, 1996–1999, 2000–2002 and 2003–2005 (SCI on CD-ROM, with correction for late processing of 2005 papers).

ISO	1996–1999	%, 1996–1999	2000–2002	%, 2000–2002	2003–2005	%, 2003–2005	Mean
Wld	12,881	100.0	15,123	100.0	17,955	100.0	15,076
EU12	4497	34.9	5123	33.9	5805	32.3	5077
US	3775	29.3	4086	27.0	4643	25.9	4129
AS5	1746	13.6	2556	16.9	3452	19.2	2501
JP	940	7.3	1150	7.6	1257	7.0	1098
DE	957	7.4	1055	7.0	1126	6.3	1037
UK	964	7.5	1041	6.9	1099	6.1	1028
FR	742	5.8	865	5.7	964	5.4	846
CN	396	3.1	750	5.0	1358	7.6	791
IT	496	3.9	611	4.0	719	4.0	598
CA	478	3.7	502	3.3	587	3.3	518
RU	486	3.8	476	3.1	432	2.4	467
ES	320	2.5	382	2.5	508	2.8	395
AU	255	2.0	284	1.9	330	1.8	286
KR	188	1.5	302	2.0	393	2.2	284
NL	253	2.0	281	1.9	326	1.8	283
IN	221	1.7	243	1.6	336	1.9	262
SE	192	1.5	231	1.5	260	1.4	224
TW	165	1.3	227	1.5	262	1.5	213
CH	165	1.3	188	1.2	236	1.3	193
PL	144	1.1	210	1.4	230	1.3	190
BR	118	0.9	187	1.2	252	1.4	179

Countries are ranked by overall mean.

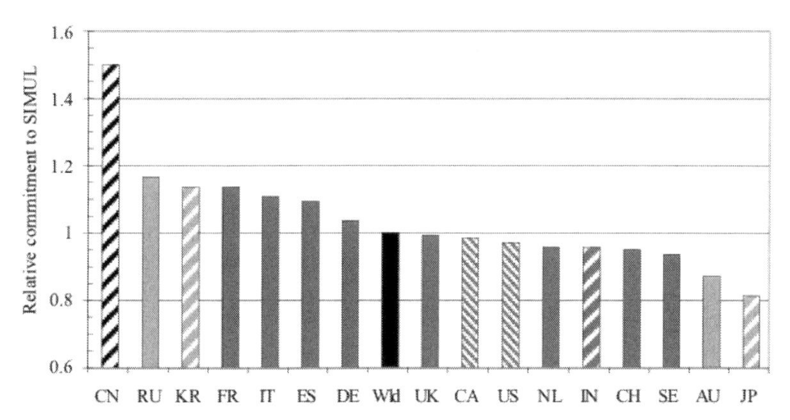

Fig. C.2. Relative commitment of some leading countries to simulation research, 1996–2005 (integer counts). Shading shows geographical region of countries.

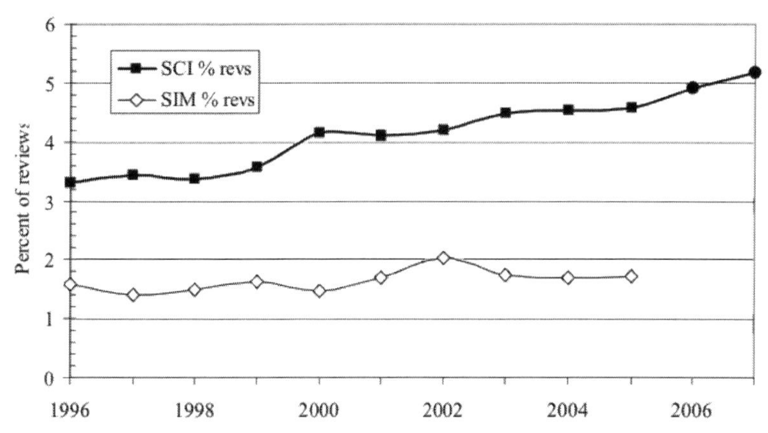

Fig. C.3. Percentage of reviews (among articles + reviews) in simulation (open diamonds, light line) and in all science (solid squares, heavy line), 1996–2007. Values taken from CD-ROMs except for science in 2006, 2007 (solid circles).

decade. The USA, closely followed by Germany, and also Spain and Russia, show to advantage in simulation, better than their positions in all science, whereas the UK's scientists are more highly esteemed in science overall than in simulation, as are the researchers in Australia. South Korea and China do not score well on this criterion.

A more conventional measure is the impact factor of the journals in which a country's papers are published. Figure C.5 shows the mean values for the last three years, 2003–2005, based on both integer and fractional

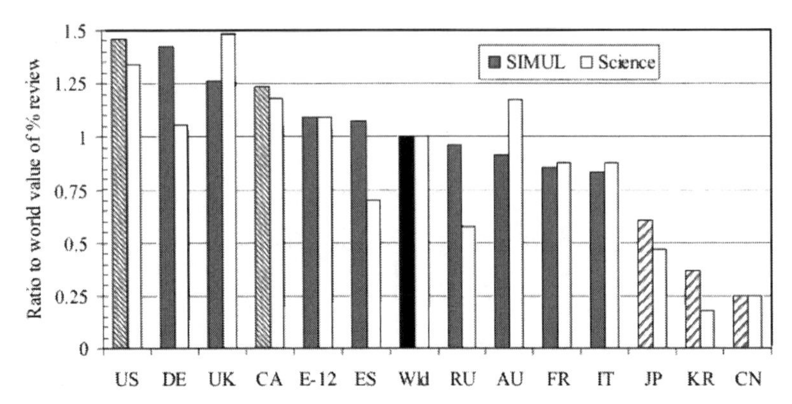

Fig. C.4. Ratios of percentage of reviews for leading countries to world values, mean of values for 1996–1999, 2000–2002 and 2003–2005, in simulation research (dark and shaded bars) and in all science (open bars). Based on integer counts.

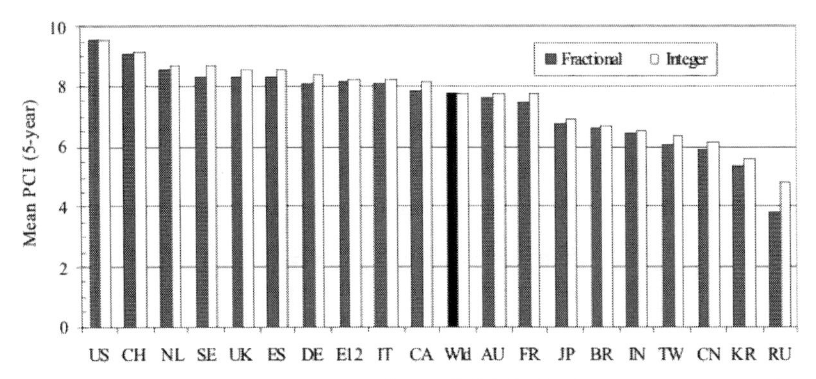

Fig. C.5. Mean journal five-year impact factor (PCI) for leading countries in simulation research, 2003–2005, fractional (solid bars) and integer counts (white bars).

counts, which give a better indication of the potential impact of countries' research, particularly small countries, where international collaboration can mask their true performance on this measure.

The USA leads also on this measure, but is closely followed by several European countries, notably Switzerland, the Netherlands and Sweden. There is a noticeable drop in journal impact factor from the level for the European countries to that for those outside Europe, notably countries in the Far East.

One of the reasons for the USA's superior performance may be that its simulation research is relatively more in those fields where papers are

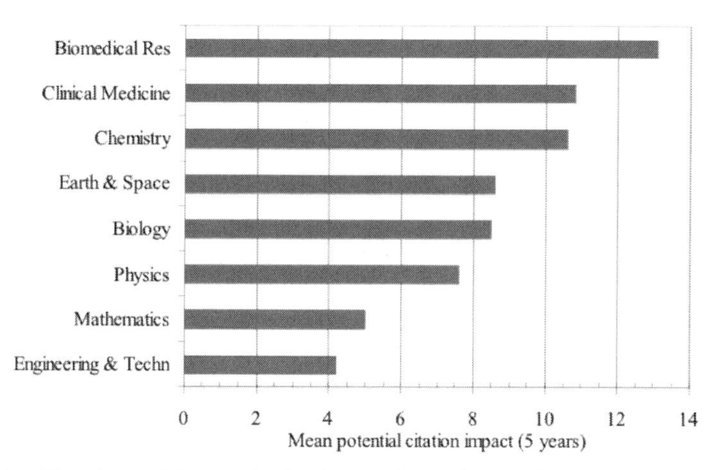

Fig. C.6. Mean journal 5-year citation impact factor for simulation papers in different fields, 2003–2005.

published in high impact journals. There are quite large differences between the fields in this respect, see Fig. C.6, with biomedical research being published in journals having more than three times as many cites, on average, as work in engineering.

Some of the differences between countries in the PCI of their papers can be explained by their relative concentration on different aspects of simulation, because, for example, China and Japan do much more engineering and much less biomedical research. Table C.9 shows the effects of this differential field pattern, on the simplistic assumption that all countries' simulation research in each field had the world mean PCI.

The differences between the mean PCI values are not large, 0.67 between the USA and China. This only accounts for one-fifth of the difference between the two countries seen in Fig. C.5, but it does show that the effects of field distribution should not be neglected when national research evaluations are being made.

The actual citation impact was determined from the WoS for papers from the leading countries for the year 2004, with citations counted over the four years, 2004–2007. This shows that the papers from some European countries, Switzerland, Denmark and the Netherlands, are actually better cited, on average, than those from the USA. However, the analysis shows that papers from the four Asian countries (Japan, India, China and South Korea) are all rather poorly cited. The numbers of papers from the

Table C.9. Distribution of simulation research between fields for the world, the USA, EUR12, China and Japan, 2003–2005, percentages, and estimates of the corresponding contribution (cont) of each field to the overall PCI (shown in bold).

	World		USA		EUR-12		China		Japan	
	%	cont	%	cont	%	cont	%	cont	%	cont
Biology	3.1	0.26	3.7	0.32	3.1	0.27	1.5	0.13	1.4	0.12
Biomed. res	7.3	0.96	9.0	1.18	7.4	0.97	4.2	0.55	5.4	0.71
Chemistry	15.1	1.60	13.6	1.44	14.4	1.52	15.9	1.68	13.2	1.40
Clinical Med.	5.6	0.61	7.5	0.81	5.8	0.62	3.0	0.32	3.9	0.42
Earth & Space	7.9	0.68	10.7	0.92	8.4	0.72	4.7	0.41	7.8	0.67
Eng & Techn	21.9	0.92	19.5	0.82	18.8	0.79	25.6	1.08	25.3	1.06
Mathematics	8.5	0.42	8.1	0.41	10.2	0.51	9.1	0.45	3.4	0.17
Physics	29.9	2.27	26.9	2.04	31.2	2.37	35.8	2.72	39.3	2.99
Others	0.7	0.07	1.0	0.09	0.7	0.06	0.3	0.02	0.3	0.03
Total	100	**7.79**	100	**8.03**	100	**7.84**	100	**7.36**	100	**7.56**

individual countries are not large, so this result should be treated with caution and would need to be confirmed for other publication years.

C.4 Results: Leading Institutions

Because of the lack of unification of institution names, a different approach was used, and in each of the selected countries (which it was thought possible that the panel might visit) the leading cities represented in the addresses of the papers from the years 2003–2005 were tabulated, on both fractional and integer counts using a special macro written by Dr. Philip Roe, and then the leading institutions within these cities were identified (not listed here). The intention here was to give additional weight to cities where there were more than one active institution working in simulation research so as to make the panel's visit program more efficient. In addition, some indication was given of the relative balance of simulation research effort between the main fields, in the form of ratios of research output to the overall mean. The purpose here was to indicate to the panel which expertise should be available from within its membership (biology, chemistry, engineering or physics) so that there would be at least one person present to take part in the discussions from the discipline in which the proposed institution for the visit was particularly strong.

Table C.10 shows the results for 44 cities in 12 western European countries and Table C.11 shows the results for 20 cities in four Far Eastern countries. The cities listed are not necessarily the ones selected for visits by the panel because of other considerations, such as the presence of particular individuals known to panel members, or simple logistics.

The listing is not perfect as it ignores the research efforts of commercial companies who may publish only occasional papers, and also those of major collaborative research centres such as CERN in Geneva, Switzerland, where the papers may have literally hundreds of addresses of which only a few will be local ones. London has been divided into several parts in Table C.10 on the basis of the postcode areas in the addresses: London SW connotes Imperial College and London WC connotes University College. If these were combined (with other postcode areas), London would easily top the list, but its output would still be much less than those of Beijing and Tokyo.

The bibliometric analysis also indicated to panel members the relative specialization of the different cities, in terms of whether the numbers of their papers in the different fields were above or below what might have been expected on the basis of the overall distribution of papers between

Table C.10. Leading cities in Western Europe, 2003–2005, in terms of their outputs of papers on simulation research.

ISO	City	Frac	Int	ISO	City	Frac	Int
FR	Paris	368	661	DE	Garching	137	243
ES	Madrid	328	497	UK	Manchester	136	218
IT	Rome	283	441	SE	Uppsala	134	231
UK	Cambridge	279	475	SE	Gothenburg	133	197
CH	Zurich	270	425	NL	Eindhoven	129	185
UK	Oxford	249	411	DE	Karlsruhe	128	204
IT	Milan	236	367	IT	Trieste	127	214
DE	Berlin	227	379	IT	Turin	127	198
ES	Barcelona	217	343	DE	Dresden	123	194
AT	Vienna	200	326	UK	Edinburgh	118	190
DE	Stuttgart	197	311	UK	Leeds	116	172
UK	London SW	192	302	BE	Ghent	113	152
UK	London WC	189	324	FR	Orsay	112	221
SE	Stockholm	188	300	DE	Aachen	111	158
NL	Delft	183	262	FR	Marseille	110	190
FR	Toulouse	182	299	DE	Darmstadt	105	153
NL	Amsterdam	176	277	BE	Brussels	104	183
FI	Helsinki	149	220	DE	Munich	104	173
FR	Grenoble	146	269	FR	Villeurbanne	103	160
CH	Lausanne	143	234	FR	Montpellier	103	187
IT	Bologna	143	220	SE	Lund	103	155
BE	Louvain	138	223	IT	Pisa	101	163

Listing in descending order by fractional counts (frac), with cities listed if they had more than 100 papers.

Table C.11. Leading cities in the far east (China, Japan, South Korea and Taiwan), 2003–2005, in Terms of their outputs of papers on simulation research.

ISO	City	Frac	Int	ISO	City	Frac	Int
CN	Beijing	916	1273	JP	Aichi	239	376
JP	Tokyo	734	1137	KR	Taejon	232	333
JP	Ibaraki	428	691	TW	Taipei	226	306
KR	Seoul	413	584	CN	Nanjing	223	349
CN	Hong-Kong	377	591	JP	Fukuoka	143	211
CN	Shanghai	369	510	TW	Hsinchu	132	175
JP	Osaka	291	451	CN	Changchun	129	163
JP	Kanagawa	273	496	CN	Dalian	127	170
JP	Miyagi	267	395	CN	Xian	126	181
JP	Kyoto	259	403	CN	Wuhan	125	183

Listing in descending order by fractional counts (Frac), with cities listed if they had more than 125 papers.

Table C.12. Relative concentration of simulation research in six fields (BIOL, biology, biomedical research and clinical medicine; CHEM, chemistry; EA&SP, earth and space; ENGR, engineering and technology; MATH, mathematics; PHYS, physics), 2003–2005, for 10 European and 6 Far East cities.

City	ISO	BIOL	CHEM	EA&SP	ENGR	MATH	PHYS
London	UK	**1.70**	0.73	1.06	0.84	1.06	0.81
Paris	FR	0.86	0.93	**1.51**	*0.63*	**1.54**	1.11
Cambridge	UK	1.22	0.99	**1.53**	*0.68*	0.74	1.06
Zurich	CH	1.25	1.28	1.36	*0.66*	1.11	0.86
Oxford	UK	**2.00**	0.74	0.91	*0.46*	0.93	1.04
Milan	IT	1.35	1.01	0.81	0.99	1.38	0.78
Berlin	DE	0.87	1.13	0.76	*0.46*	1.30	1.39
Barcelona	ES	0.75	**1.70**	1.05	*0.60*	1.23	1.00
Stuttgart	DE	*0.28*	0.87	*0.32*	1.31	**1.63**	1.25
Amsterdam	NL	**1.52**	**1.49**	0.90	*0.33*	1.05	0.94
Beijing	CN	*0.59*	0.76	1.01	1.21	1.03	1.19
Tokyo	JP	0.71	*0.65*	**1.42**	1.06	*0.47*	1.33
Ibaraki	JP	*0.53*	0.76	1.12	1.01	*0.12*	**1.60**
Seoul	KR	*0.62*	*0.69*	1.25	1.36	0.83	1.09
Hong Kong	CN	0.74	*0.50*	0.83	**1.59**	**1.67**	0.82
Shanghai	CN	*0.57*	0.93	*0.39*	**1.42**	1.09	1.10

Values over 1.41 shown in bold; values below 0.71 shown in italics.

fields, see Fig. C.6 and Table C.9. The results are shown in Table C.12 for ten leading European cities and six in the Far East.

Again, this listing is not necessarily exact because company outputs may well not be fully represented, and a city may contain several institutions whose outputs may differ from the average for the locality. But it does suggest what expertise should be available within sub-panels, e.g., biology/medicine for Amsterdam, London and Oxford, chemistry for Amsterdam and Barcelona and physics for Ibaraki and Berlin.

C.5 Conclusions

The main lessons from this analysis are as follows:

- Simulation represents (after correction for the lack of recall of the filter) about 5% of the papers in the SCI, but as a subject it is growing relatively quickly, at about 5% per year compared with 2.5% for science as a whole (Fig. C.1).
- Within simulation research, the country with the largest output of papers is the United States, with a fractional count of 26% of world

papers in 2003–2005. This is less than that of 12 leading western European countries combined (32%). Both outputs are declining as a percentage presence in the world, but are increasing in absolute amount (Table C.8).

- China's output was second highest, reaching 13% of the world total (integer counts) in 2007, followed by that of Germany and the UK (8%), France (7%) and Japan (6%).
- China also had the highest relative commitment to simulation research over the decade 1996–2005, with 50% more papers than expected on the basis of its overall output. The US published about 10% fewer papers than expected on this basis (Fig. C.2).
- Three measures of esteem were determined: percentage of reviews, potential and actual citation impact. The USA was in first place on the first of these (Fig. C.4), followed by Germany and the UK. It was also in first place for PCI (Fig. C.5), followed by Switzerland, but only in fourth place for actual citations to 2004 papers (Fig. C.7).
- The Far East countries all scored relatively poorly on each of these three indicators; this was partly because they concentrated their efforts in fields where the papers were in low impact journals and received few citations (such as mathematics and engineering, compared to biomedical research; Fig. C.6).

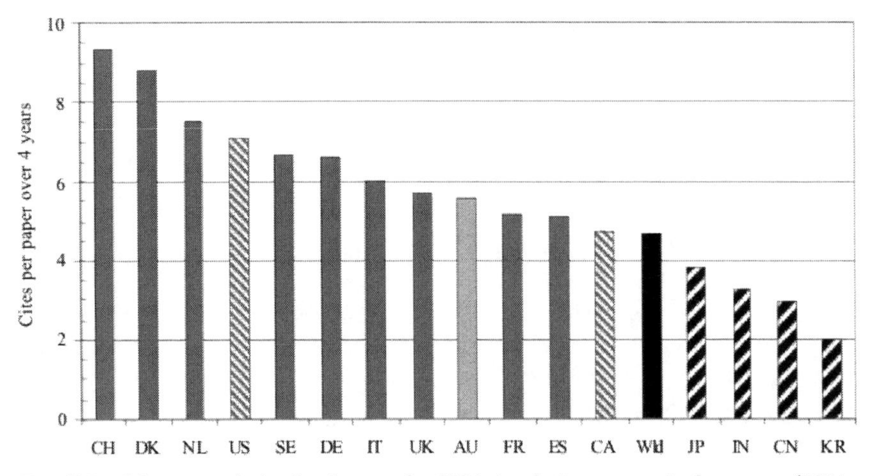

Fig. C.7. Mean actual citation impact for 2004 simulation papers in four years (2004–2007) from leading countries.

- The tally of papers for 2003–2005 enabled lists of the leading cities in western Europe and the Far East to be produced (Tables C.10, C.11) for the information of the panel members when they were deciding which sites to visit. The relative specialization of these cities in different scientific fields was also determined (Table C.12) in order to help the panel to arrange appropriate sub-panel membership for its visit program.

Appendix D

GLOSSARY

211	Trans-centenary key construction in higher education projects (P.R. China)
863	High-tech program/project (P.R. China)
985	Projects focused on building world-class universities in the 21st century (P.R. China)
3D-RISM	Three-dimensional reference interaction site model of Japan's IMS
ADME/Tox	Absorption, distribution, metabolism, excretion, and toxicity (assay; screening)
AIAA	American Institute of Aeronautics and Astronautics
Algorithm	A sequence of instructions, often used for calculation and data processing; a method in which a list of well-defined instructions for completing a task will, when given an initial state, proceed through a well-defined series of successive states, eventually terminating in an end-state.
ASCE	American Society of Civil Engineers
BABE	Burn at both ends
CAE	Computer-assisted engineering
CAE	Chinese Academy of Engineering (PRC)
CAPE	Computer-aided process engineering
CAS	Chinese Academy of Sciences (PRC)
CBS	Center for Biological Sequence Analysis at the Technical University of Denmark (DTU)
CCP	Collaborative computational projects (United Kingdom)
CFD	Computational fluid dynamics
CFD/FEM	Computational fluid dynamics/finite element method
CNOOC	China National Offshore Oil Corporation
COE	Centers of excellence (Japan, MEXT program/University of Tokyo)

CoW	Circle of Willis (a ring-like arterial structure sitting at the base of the brain whose main function is to evenly distribute oxygen-rich arterial blood to the cerebral mass)
CPU	Central processing unit
CRDL	(Toyota) Central R&D Labs, Inc.
CREST	Core research for evolutional science and technology program of the Japan Science and Technology Agency, JST
CSL	Computational Science Laboratory (Mitsubishi)
CSE	Computational science and engineering
CSTP	Council for the Science and Technology Policy (Japan)
CT	Computed tomographic
DFG	German Research Foundation (Deutsche Forschungsgemeinschaft)
DFT-GGA	Density functional theory-generalized gradient approximation
DG	Distributed generation
DMFT	Density matrix functional theory (also, dynamical mean field theory)
DPD	Dynamic panel data
DSA-CT	Digital subtraction angiography-computed tomography
EBI	Energy Biosciences Institute
EEG	Electroencephalography
EMSL	Environmental Molecular Science Laboratory at the Department of Energy's (DOE's) Pacific Northwest National Laboratory (PNNL)
EPRI	Electric Power Research Institute
ESC	Earth Simulator Center of the Japan Agency for Marine-Earth Science and Technology
FE	Finite element
FEAP	Finite element analysis program
FEM	Finite element method
flops	(or FLOPS or flop/s) Floating point operations per second
FMO	Fragment molecular orbital
fMRI	Functional magnetic resonance imaging
FPGA	Field-programmable gate array
FPMD	First principles molecular dynamics
GGA DFT	Generalized gradient approximation density functional theory
GPU	Graphic processing unit

GSIC	Global Scientific Information and Computing Center at the Tokyo Institute of Technology
gTOW	Genetic tug of war
Hamiltonian	In quantum mechanics, the Hamiltonian H is the observable corresponding to the total energy of the system; ... the set of possible outcomes when one measures the total energy of a system. (http://wikipedia.com)
HEP	High-energy physics
HPC	High-performance computing
HPRC	High-performance reconfigurable computing
ICME	Integrated computational materials engineering
IEEE	Institute of Electrical and Electronics Engineers
IP	Intellectual property
IRMOF	Isoreticular metal organic framework
IUPS	International Union of Physiological Sciences
IWM	Fraunhofer Institute for Mechanics of Materials (Institut Werkstoffmechanik)
JLPSM	Joint Laboratory of Polymer Science and Materials of the Institute of Chemistry, Chinese Academy of Sciences (ICCAS)
LANL	Los Alamos National Laboratory
LBM	Lattice-Boltzmann method
LES	Large eddy simulation
LIDAR	Light detection and ranging
Linpack	A collection of FORTRAN subroutines that analyzes and solves linear equations and linear least-squares problems.
LLNL	Lawrence Livermore National Laboratory
MCC	Mitsubishi Chemical Corporation
MCHC	Mitsubishi Chemical Holdings Company
MCRC	Mitsubishi Chemical Group Science and Technology Research Center
MD	Molecular dynamics
MEMS	Microelectromechanical systems
METI	Ministry of Economy, Trade, and Industry of Japan
MEXT	Ministry of Education, Culture, Sports, Science, and Technology of Japan
MLWF	Maximally localized Wannier function
MM	Molecular mechanics

MOF	Metal organic framework
MOST	Ministry of Science and Technology (P.R. China)
MOU	Memorandum of understanding
MPI	Message-passing interface parallel programming standard, the de facto industry standard
MRA	Magnetic resonance angiography
NAREGI	National Research Grid Initiative (Japan)
NCAR	National Center of Atmospheric Research (Japan?)
NCSA	National Center for Supercomputing Applications at the University of Illinois at Urbana-Champaign
NEDO	New Energy Development Organization (Japan)
NEMS	Nanoelectromechanical systems
NGSP	Next Generation Supercomputing Project (Japan/Riken/IMS)
NINS	National Institutes of Natural Sciences (Japan)
NIST	National Institute of Standards and Technology
NMR	Nuclear magnetic resonance
NNSA	National Nuclear Security Administration
ODE	Ordinary differential equation
OECD	Organization for Economic Co-operation and Development
OGSA	Open grid services architecture
OLED	Organic light-emitting diode
ONIOM	Our own N-layered integrated molecular orbital and molecular mechanics
PAYAO	Web 2.0 community tagging system
PDE	Partial differential equations
PDF	Probability density function (distribution)
PET	Photoinduced electron transfer
PGAS	Partitioned global address space programming languages
PHG	Parallel hierarchical grid
PI	Principal investigator
PMS	Process modeling and simulation
PPD	Parallel partitioning diagonal
PRACE	Partnership for advanced computing in Europe
QCD	Quantum chromodynamics
QM	Quantum mechanical
RLC	(circuitry consisting of) Resistive, inductive, and capacitive (elements)
RMB	Ren min bi (the currency of the People's Republic of China)

RRPS	Ratio of reduction in power supply
SAC	Symmetry adapted cluster
SAC-CI	SAC-configuration interaction
SBGN	Systems biology graphical notation
SBI	Systems Biology Institute (in the Department of Systems Biology at the Japan Foundation for Cancer Research in Tokyo, Japan)
SBML	Systems biology markup language
SBW	Systems biology workbench
SCF	Self-consistent field
SC-PRISM	Self-consistent polymer reference interaction site model
SIMS	Secondary ion mass spectrometry
SNL	Sandia National Laboratory
SOC	Self-organized criticality
SPDE	Stochastic partial differential equation
Speedup	In parallel computing, speedup refers to how much a parallel algorithm is faster than a corresponding sequential algorithm. Linear speedup or ideal speedup is obtained when $S_p = p$ [p is the number of processors]; that is, ideally, when running an algorithm with linear speedup, doubling the number of processors doubles the speed; this is considered very good scalability.
super SINET	Super science information network
TCRDL	Toyota Central R&D Labs, Inc
Tflops	One trillion floating point operations per second
TSUBAME	Tokyo tech supercomputer and ubiquitously accessible mass storage environment
UPC	Unified parallel C (programming language)
UQ	Uncertainty quantification
V&V	Validation and verification
WSRF	Web services resource framework